A Time for Truth

A TIME FOR
TRUTH

———— ★ ————

Reigniting the Promise of America

TED CRUZ

BROADSIDE BOOKS
An Imprint of HarperCollinsPublishers

HarperCollins books may be purchased for educational, business, or sales pro-
motional use. For information, please e-mail the Special Markets Department
at SPsales@harpercollins.com.

Broadside Books™ and the Broadside logo are trademarks of HarperCollins
Publishers.

FIRST EDITION

Designed by William Ruoto

Library of Congress Cataloging-in-Publication Data has been applied for.

ISBN: 978-0-06-236561-3

15 16 17 18 19   DIX/RRD   10 9 8 7 6 5 4 3 2 1

*This book is dedicated to my parents, for their inspiration,
strength of character, and unconditional love.*

*It is dedicated to my wife Heidi—my best friend
in all the world—and our precious daughters Caroline
and Catherine, who are the joys of our life.*

*And it is dedicated to truth-tellers and freedom fighters,
to the grassroots and the courageous conservatives
whose passion and leadership are
turning our great nation around.*

CONTENTS

———————— ★ ————————

Mendacity

Pandemonium ensued. There were angry glares, heated accusations. Red-faced name-calling echoed off the walls and vaulted ceilings in a room just off the main corridor of the U.S. Capitol. It was Tuesday, February 11, 2014. Another lunch of the Senate Republicans.

I'd been a regular part of these gatherings ever since I was elected to the U.S. Senate in 2012. Most, if not all, of the then forty-five members of the Republican conference usually attended; these were, literally, free lunches after all, in some of the most beautiful rooms in the U.S. Capitol. On Tuesdays and Thursdays, we met in the Lyndon B. Johnson Room, an expansive chamber adorned with ceiling frescoes by the Italian artist Constantino Brumidi, a large gilded mirror, an opulent chandelier, and marble-paneled walls. On Wednesdays we met in the Mike Mansfield Room, a wood-paneled rectangular conference room named for the late Senate majority leader from Montana.

Typically the party lunches were civil discussions—somewhat plodding, and occasionally instructive. On this day, however, civility was not on the menu.

At this lunch, the duly elected members of the U.S. Senate—many who'd served in the august body for decades—were yelling. Not simply raising their voices or speaking loudly, but angrily yelling at their colleagues in the room who had committed what I had quickly come to learn was the cardinal sin of Washington, D.C.: telling the truth.

The events of that week provided yet another example of just how bad things in our nation's capital had become.

The issue at hand was the federal debt ceiling. Periodically it fell to the U.S. Congress to vote to raise the amount of debt the federal government can accumulate in order to continue its current spending levels.

As a U.S. senator, Democrat Barack Obama voted repeatedly against President George W. Bush's efforts to raise the debt ceiling, calling the need for such a vote a sign of "failed leadership."[1] In 2006, Obama had declared that "increasing America's debt weakens us domestically and internationally." He added at one point that "Washington is shifting the burden of bad choices today onto the backs of our children and grandchildren. America has a debt problem and a failure of leadership. Americans deserve better."[2]

This seemingly principled position changed dramatically when Obama won the presidency. This sadly is not surprising in Washington, D.C.—a place where principles are fungible, often lasting only until the next election.

Not only did President Obama abandon Senator Obama's position on the debt. He made the situation far worse. Never in the history of our country have we had a spender like Barack Obama. When the president took office, our debt was just over $10 trillion, itself a sizable figure; today the debt is over $18 trillion. Just think about that for a moment—it took forty-three presidents nearly 220 years to accumulate $10 trillion in debt.* In just six years, President Obama has almost doubled that. Our total debt is now larger than our entire economy. Today roughly 40 cents of every dollar that the federal government spends is borrowed money, which we will have to repay for years to come.[3]

In the early months of 2014, President Obama was urging Democrats and Republicans to pass yet another increase of the debt ceiling, so we could spend even more. The president was demanding from Congress what he called a "clean" bill. In the backwards parlance of Washington, the definition of "clean" was adding trillions more in debt without including any reforms to arrest Washington's out-of-control spending. That didn't seem very clean to me.

President Obama needed Congress's help to carry out his spending plans. I saw this as an opportunity. Historically the vote on the debt ceiling has proven to be one of the few tools that the U.S. Congress has been able to use to achieve any modicum of success reining in the size and power of the federal government. In the previous 55 times that Congress had raised the debt ceiling, it attached significant conditions to the legislation 28 times. In 1985, for example, Congress conditioned a debt-ceiling increase on the passage of the Gramm-Rudman-Hollings Act, one of the most constructive bipartisan efforts to rein in federal spending in modern times. In 2010, Congress used its leverage to pass the so-called Budget Control Act—which both parties touted as a serious effort to reduce federal spending. (In reality, the BCA didn't actually "cut" spending; it simply slowed its growth. And Congress has already abandoned some of those modest spending restraints.)

---

*America has continuously held a public debt since 1789, with the exception of a single year, 1835.

For months leading up to this moment, Republican leaders had pledged to their constituents that when it came time to raise the federal debt limit, they would demand meaningful spending reforms from this president. Rest assured, we were told, the Republicans would insist on it! Those of us who had fought so hard to stop Obamacare the previous year had been told, by these wizened D.C. insiders, that we had picked the wrong fight; the real fight should be over the debt ceiling.

Indeed, just days before our combative February lunch, Senate Republican leader Mitch McConnell, appearing on *Fox News Sunday*, had pledged: "I think for the president to ask for a clean debt ceiling, when we have a debt the size of our economy is irresponsible. So, we ought to discuss adding something to his request to raise the debt ceiling that does something about the debt or produces at least something positive for our country."[4] Given those public commitments, it would have been natural to expect that our lunch discussion that day would have focused on these positive "somethings" the voters were promised, to start us on a track back to fiscal sanity.

That hadn't happened. Just a week earlier, the GOP leadership in the House of Representatives had buckled. They had joined with 193 Democrats to run over 199 House Republicans and give President Obama the "clean" debt increase that he had demanded. Sadly, Senate Republican leaders wanted us to do the exact same thing.

This wasn't a total shock to me. I had figured all along that our GOP leadership, like so many times before, would offer some half-hearted proposal to deal with spending and then, under pressure, eventually surrender to the Democrats. Caving in to the president's demands had been our modus operandi for far too long. But I never thought that surrendering would be our *starting* position.

But it was worse than that. Much worse.

The U.S. Senate cherishes its myriad rules, traditions, and protocols. This sometimes produces great frustration to many of us trying to get something done, since many of these rules are vestiges of a bygone era. But this time the rules played to our advantage. For decades, the ordinary procedure in the Senate has been that in order to move to proceed to take up a debt-ceiling increase, 60 senators must vote in favor of the motion. At the time, the Senate had 55 Democrats, which meant that 5 Republicans would have to support taking up the vote. That gave our side significant bargaining power with the Democrats and the White House.

Obviously the Democrats didn't want that. But neither did the Republican Senate leadership.

In the Senate, any rule can be changed by unanimous consent, which takes, as the name implies, the affirmative consent of all one hundred senators. And so, as our lunch began, the members of the Republican leadership stood before us and asked every senator to join with the Democrats in granting unanimous consent to lower the 60-vote threshold to take up the debt ceiling to just 50 votes.

None of us should oppose this, we were told, and for two reasons. First, if we lowered the threshold, then the "clean" debt ceiling would pass, and that was very much the outcome the leadership assumed each of us really desired.

And second, if we consented to lowering the threshold, Democrats would then have the votes to raise the debt ceiling on their own. We could all vote no. This way, we could return home and tell the voters that we had opposed raising the debt ceiling, *right after consenting to let it happen.*

This time I was stunned by the chicanery, expressed openly, if not proudly, to the rest of us. Looking around the Lyndon Johnson Room, named for one of the biggest-spending presidents in American history, I had a new appreciation for why we were gathered here, beneath his smiling portrait. Most senators seemed perfectly fine with the leadership's proposal. There were nods and murmurs of assent.*

It was too much. I raised my hand and said, "There's no universe in which I can consent to that."

I had spent two years promising Texans that if they voted for me I would fight with every breath in my body to stop the out-of-control spending and debt that is bankrupting our kids and grandkids. I explained, "If I were to affirmatively consent to making it easier for Democratic Senate leader Harry Reid to add trillions in debt—with no spending reforms whatsoever—I think it would be dishonest and unfaithful to the voters who elected me."

This was an obvious point. Every one of these senators had promised their constituents that they'd oppose tricks like this to add to our debt. But as it happened, the only other senator in the room who spoke up to agree with me was Mike Lee, the junior senator from Utah and a staunch fiscal conservative. Lee, my closest friend in the Senate, was another troublemaker in the eyes of Republican leaders.

In the two years I've been in the Senate, nothing I have said or done has engendered more venom and animosity from my fellow Republicans than the simple objection I made that afternoon. Indeed, the issue coalesced the rage that career politicians already felt for upstarts like me.

That was when the yelling began or, what outlets such as *National Review* more politely called "a spirited exchange between [Cruz], McConnell, and others."⁵

---

*During my time in the Senate, I've been amazed how many senators pose one way in public—as fiscal conservatives or staunch Tea Party supporters, for example—and then in private do little or nothing to advance those principles. Indeed, if transcripts of our Senate lunches were released to the public, I think many voters would be astonished.

When I made my case to my colleagues, they looked at me like I was a fool. I heard more than one variation of "That's what you say to folks back home. You don't actually do it." They were convinced that they had a brilliant maneuver to increase our debt without any fingerprints. And here was a freshman senator with the temerity to screw it all up.

Don't you understand what we are doing? senators thundered. *Why are you forcing us to tackle this? Why can't you just go along?*

One of the saddest aspects of Washington is that when you stand for principle, and actually seem to mean it, the typical response is "What are you *really* up to?" Almost nobody just tells the truth in Washington. There's always an ulterior motive. You're trying to raise money. You're building an email list. You're preening for the cameras. You're running for president. Anything other than that you're doing what you're doing because you believe in every fiber of your being it's the right thing.

The yelling during our lunches continued over the entire week.

Sitting through the harangues wasn't terribly difficult, as long as I remembered that I didn't work for the Senate leadership but for 26 million Texans. Frankly, I've found the more reviled you are in Washington, the more they appreciate you in places like Waco, and Dallas, and San Antonio.

Still, it was surreal to see Republican senators—some of who were conservative icons—turn purple over this issue. The fulminations went on and on against me and my crazy idea that senators ought to vote "yes" on what they support and "no" on what they oppose.

Over time it occurred to me that what was prompting the greatest anger was not that I opposed raising the debt ceiling. What infuriated them was that by objecting to the unanimous consent deal they had cut, it forced those Republicans who wanted the debt ceiling raised to actually admit so in public with their recorded vote. It prevented them from misleading their constituents.

It's often suggested that constitutional conservatives dislike compromise. That's the attack the media likes to wage on folks on the right—even though to them compromise is just giving the Democrats whatever they are asking for. In fact, I'm happy to compromise—with Democrats, Republicans, Independents, or Libertarians (heck, I've joked I'll compromise with Martians!)—*if* we are actually shrinking the size, power, or spending of the government. *If* we are expanding individual liberty or defending the Constitution.

I tried to make the case that if we all stood together, we could force the Democrats to compromise for once. To work with us on spending restraints. Just as Republicans and Democrats had done with different presidents in the past. Just as Barack Obama had once insisted that Democrats do with the Bush administration. I wasn't looking to pass a flat tax or a sweeping proposal to rein in entitlement spending once and for all. We were a minority party in the Senate, and John Boehner had already caved in the House. It was obvious we couldn't get everything we wanted. But was the alternative abject surrender?

My position was not that we should never raise the debt ceiling. That's not a responsible position, unless you are prepared to cut the federal budget by 40 percent overnight (which no one has proposed a sensible path to doing). So I offered the conference a few concrete, relatively modest suggestions for what Republicans could get out of the debt ceiling. For one, I argued that Republicans could insist on attaching legislation to the debt ceiling bill that prohibited bailing out big insurance companies under Obamacare. This would address one of the biggest sources of our looming fiscal crisis. Under the current law, Obamacare automatically funnels billions of tax dollars to bail out giant insurance companies. That was how the Obama administration won those companies' support. And it would be very hard for Senate Democrats to fight. How many Democrats, in an election year, would want to die on the barricades to defend insurance company bailouts?

A second possibility was, admittedly, a bigger reach—requiring the beginnings of entitlement reform, to make modest fixes to those vital programs in order to make them more fiscally sustainable going forward. Harder, but very meaningful if we could get it done.

A third was attaching Pennsylvania senator Pat Toomey's "Full Faith and Credit Act," or what I call the Default Prevention Act. It addressed the bogeyman that President Obama puts forward—that we will default on our debt if the debt ceiling is not raised. Default, according to Obama, would result in a worldwide financial apocalypse.

Nobody wants a default. And even if the debt ceiling were not raised, no responsible president would allow a default. Each month, federal revenues are about $200 billion; federal debt payments are $30–40 billion. There is ample revenue to always, always, always pay our debts (assuming the president doesn't try to scare the markets by threatening to force a default). So Senator Toomey's bill expressly required the United States to always pay its debt no matter what, even if the debt ceiling isn't raised. Even as good a demagogue as the president is, and even as terrible as Republicans typically are at communicating, it would be hard to say that Congress was risking default of the debt in order to pass legislation that would in fact make it *impossible* for us to ever default on our debt.

I urged other Republican senators to make their own suggestions. None was forthcoming. The response from the Republican leadership was firm—they didn't want us to fight for anything. The House had surrendered, and we had to do so as well.

In exchange for another increase in the debt ceiling, they wanted *nothing*. They feared accusations that Republicans were willing to risk a default of the U.S. government. They feared bad press. They feared Obama. They feared pushing for anything that was hard. The great irony, of course, was that opposing an increase in the federal debt was an 80-20 issue in most polls. If we explained to the American people what we were doing—if we were honest with them instead of trying to trick them—I believed we could win the argument.

But it wasn't about solving the debt. It was about self-preservation. One senator asked me directly, *"Why do you want to throw five Republicans under the bus?"*

"I don't," I replied. "It's leadership that is asking you to jump under a bus. Every single one of you campaigned telling your constituents the same thing I did; all I'm asking is that you actually do what you told the voters you would do."

The vote to move forward on the debt ceiling came the following day. Because Mike Lee and I had refused to change the rules, 60 votes were still needed. Ordinarily in the Senate, these votes take fifteen minutes. For this particular vote, fifteen minutes came and went. And there were only 59 "ayes"; there weren't enough votes to give Obama the "clean" bill he wanted. I was hopeful that maybe Republicans would stand firm behind their convictions.

But something peculiar happened. Traditionally, when a senator casts a vote, the clerk of the Senate reads it aloud—either "Mr. Smith, aye," or "Ms. Jones, nay."

For this particular vote, however, the clerk was instructed not to read the votes aloud as they were cast.[6] That meant the votes could be kept secret, so that the leadership could work on getting enough Republicans who voted against the debt ceiling motion to quietly switch their votes without anyone knowing.

The fifteen-minute vote stretched to an hour. Only four Republicans had voted yes; to win, the Democrats needed one more. Under ordinary rules, the Democrats had lost. But they didn't look like losers to me. The Democrats were sitting back and smiling while our GOP leaders went out of their way to help Harry Reid and President Obama pass yet another increase to our massive federal debt.

Finally, after spending an hour leaning heavily on Republicans to switch their votes, the top two Republican leaders walked to the clerk and simultaneously voted in favor of the motion. Six other Republicans then followed suit, switching from "no" to "yes."

All told, twelve Republicans voted with the Democrats. The remaining Republican senators, 33 of them, a majority of our conference, stood against out-of-control spending and the Obama administration. None of that mattered, however. The final result was that President Obama, Harry Reid, and House Democratic leader Nancy Pelosi, with the active complicity of the Republican leadership, were able to add trillions of dollars to the national debt—a bill that we are passing along to our children and grandchildren—while doing nothing whatsoever to control future spending.

We lost the battle over the debt ceiling, but that was just one skirmish in the fight to make D.C. listen.

Outside the Beltway, it is obvious to everybody that Washington is broken, that it's not committed to solving the real problems we face. No matter what politicians say on the stump, it seems that when they get to Washington they almost always forget the people who elected them.

All across America, you hear the same thing, from Republicans, Democrats, Independents, and Libertarians: *Our elected officials aren't listening to us.*

Regardless of which party is in power, government keeps growing, the lives of hardworking Americans keep getting harder, and the disconnect between the government and the American citizenry keeps getting greater.

The driving force in Washington on the left and the right is risk aversion. Folks in our nation's capital perpetually grow the size and power of the government because that's the easiest thing to do. Fighting for principle, fighting against a powerful media and even sometimes against popular big-spending proposals—well, that's a lot harder. Far too much decision making revolves around lobbyists and the entrenched political powers, which means that the only way to enact meaningful change is to fundamentally change the rules.

In another context, the Silicon Valley tech community has seen this same fight repeatedly. Over and over again, we've seen disruptive apps that change the means of delivery of a good or service. And, inevitably, the existing providers fight back. They don't want change. So, when an Uber or Lyft comes into a city, the incumbent taxi commissions inevitably fight tooth and nail to maintain the status quo.

Telling the truth in Washington, D.C., is a radical act. And it earns you the enmity of career politicians in both parties. When you tell the truth about Washington—when you expose the fact that elected officials are misleading the voters who elected them—you pay a price. It's one thing to criticize the other party. But when you admit publicly that many of those in your own party are complicit in the problem, well, that's when the long knives come out.

The leadership in Congress has two ways of punishing members who speak the truth. First, they go after the "lifeblood" of politics: money. Without campaign money, you cannot get elected, and you cannot get your message out. Especially in this modern media age, communicating is expensive. And so is connecting with and mobilizing the grassroots.

The first year that I was serving in the Senate, we saw considerable campaign donations come in from the corporate world; Texas is, after all, home to more Fortune 500 companies than any other state. Once I demonstrated that I would stand up to party leadership, that stopped—cold. Members of the GOP leadership made it clear to K Street lobbyists and political action committees that if they continued to support me, then they would be frozen out. Our PAC fund-raiser quit, under pressure from folks connected to the GOP leadership. They did the same thing to Mike Lee.

Want to know why so many elected officials listen to party bosses instead of their constituents? Washington is corrupt, and control over D.C. money is a big part of it.

Fortunately, my campaign was never dependent on financial support from giant corporations or K Street lobbyists. When I ran for Senate, almost all of them opposed me; after I was elected, we welcomed their support, but I never forgot for whom I was working.

As a grassroots candidate, our campaign contributions came instead from tens of thousands of citizens in all fifty states. They were young people, small business owners, little old ladies sending in five or ten dollars so that we could stand together and change Washington. They weren't susceptible to threatening phone calls from leadership aides. Cutting off our money wasn't going to deter me from speaking the truth.

The second method of punishment the GOP elites use is public flogging. Anonymous quotes appear in Capitol Hill publications from unnamed Republican sources—they're usually Republican leadership staff members—wielding nasty personal insults.

"*[Democratic candidate for governor of Texas] Wendy Davis has more balls than Ted Cruz.*"

"*Ted Cruz came here to throw bombs and fund-raise off of attacks on fellow Republicans. He's a joke, plain and simple.*"

"*He's an amateur.*" "*A fraud.*" "*A hypocrite.*" "*A wacko bird.*"

"Jim DeMint without the charm."

All of those things were said by *Republicans*. And most of them were attributed to anonymous "senior GOP aides."

The Republican leadership's attacks are amplified and made more effective by using friendly media outlets. When leadership is displeased, they place hit pieces with journalists only too happy to cooperate. Indeed, so much so that one particular *Politico* reporter often seems like he is Mitch McConnell's press secretary; nearly every attack from leadership gets echoed and amplified in his stories. As this reporter noted after one Senate lunch (apparently without irony): "The closed-door [Senate lunches] are supposed to be private . . . so senators interviewed for this article asked not to be named."[7]

Of all the friendly media outlets for GOP leadership, none is more potent than the editorial page of the *Wall Street Journal*. That page has built a remarkable legacy of defending free-market ideals and standing up to government power. In the 1970s and '80s, legendary editor Robert Bartley helped set the stage for the many successes of the supply-side economics of Jude Wanniski, Jack Kemp, and Ronald Reagan.

Whenever congressional leadership is particularly exercised on a topic, it usually takes about seventy-two hours for the *Journal* editorial page to unleash that same attack in the most important space in journalism.

Thus, a couple of days—like clockwork—after the battle on the debt ceiling, the *Journal* penned a blistering editorial titled "The Minority Maker." Now, one might think that the *Wall Street Journal* would be concerned about our $18 trillion debt, and dismayed that Congress had not even tried to check the administration's profligate spending. But no, the *Journal*'s concern was different, namely that my fighting on the debt ceiling had made Republican leadership look bad.

Mirroring the words of Senate Republican leaders, the *Journal* intoned: "We're all for holding politicians accountable with votes on substantive issues, but Mr. Cruz knew he couldn't stop a debt increase the House had already passed. He also had no alternative strategy if the bill had failed, other than to shut down the government again, take public attention away from ObamaCare, and make Republicans even more unpopular."[8] (As history showed, the editorial's central premise—that the fight on the debt ceiling would make it far more likely Harry Reid and the Democrats would keep control of the Senate—was somewhat in conflict with the actual results of the 2014 election.)

That was the first shot. The next day, our office received a call from an editorial writer at the *Journal* who was writing a follow-up piece on the very same topic.

I spent more than an hour on the phone with the writer, arguing that any effort to use the debt ceiling to force spending reforms was far from useless. We could get concessions from Obama, as Mitch McConnell and others had asserted only days earlier. I had no desire to tilt at windmills, I explained on the call, but their premise—that we could never succeed—was belied by the facts. In some sense, I understood where the *Journal* was coming from. They thought my tactics were myopic, as the GOP leadership had no doubt convinced them. But I observed that any responsible journalist, even if he or she thought my position was a political blunder, would have to at least acknowledge that the debt ceiling has historically been the best leverage that Congress has on spending; indeed, Congress had attached meaningful conditions to 28 of the last 55 debt ceiling increases. It had proven true even with President Obama and Harry Reid. Just one year earlier they had been forced to agree to the Budget Control Act through the leverage of the debt ceiling.

The crowning moment of our telephone exchange was when the writer asserted that every single Republican strongly opposed raising the debt ceiling without any spending reforms. My response seemed to baffle her: "If you lead the fight to pass something . . . then you don't oppose it."

Alas, the conversation proved fruitless. Neither of the *Journal's* editorial pieces had any acknowledgment of the historical efficacy of using the debt ceiling for leverage. Instead, they simply asserted that resistance was futile.

As intended, this double hit in the press had its effect—especially in the business and donor community. It was here where I made a mistake—I didn't realize how effective the media attacks, especially in the conservative media, would be. Over and over again, I had to explain myself to people who usually were on our side, especially because the attacks from supposed Republican allies often inverted what was actually happening. For example, just after the debt ceiling fight, Bob Woodward went on one of the Sunday shows and said, "Mitch McConnell . . . thinks that Cruz is literally the most selfish senator he's ever seen." There was almost a Freudian projection at work here. It seems to me that the "selfish" thing for me to have done would have been to go along to get along: go to the Washington parties, raise money, stay in office for life, and never rock the boat. Who in their right mind would "selfishly" choose to endure the vilification, the personal and nasty attacks? In the curious lexicon of Washington, what is perceived as "selfish" is doing what you said you would do: honoring your commitments to your constituents. Because, I suppose, that selfishly makes it uncomfortable for others who do not wish to do the same.

The typical Washington response would be for me to fight back, to fight fire with fire. To insult my colleagues, to plant nasty stories and give anonymous quotes smearing those on the other side. But I'm not willing to play that game. It's wrong, and it's not what Texans elected me to do. Instead, when the attacks come my way, I have not reciprocated. What I have done and what I intend to continue to do is speak the truth and explain how money and power in Washington conspires to create even more money and power for the politically connected at the expense of the hardworking taxpayers.

If I had it to do over again, I would have made a bigger effort to inform my supporters and impartial journalists of what actually happened. But I foolishly thought the facts were so abundantly clear that the media would accurately describe what was happening.

As I reflected on the situation, I was struck by the fact that many of the weapons used against me—mischaracterizations of motivations, attacks in the press, efforts at ostracism—are very similar to how many Washington Democrats and Republicans have greeted the tea party, the millions of conservative activists who rose up in response to the out-of-control spending, debt, and government power in the Obama era.

Many reporters like to discuss the Republican Party as currently divided between the tea party and the establishment. But that line misunderstands what is actually going on. The biggest divide in politics is not between the tea party and the establishment. It is not even between Republicans and Democrats. The biggest divide is between career politicians—in both parties—and the American people.

There is a fundamental corruption in Washington—those in power get fatter and happier while the rest of America suffers as a consequence of their policies. I've seen it firsthand. The counties surrounding the nation's capital—where most legislators and lobbyists live—are soaring in median income, while the median income for American families has stagnated nationwide. Indeed, five of the six wealthiest counties in the entire country are suburbs of Washington.

In the Roaring Twenties, Calvin Coolidge memorably observed, "the chief business of the American people is business." Well, the chief business of Washington is government, and in today's climate, business is good. But, in the rest of the country, we have the lowest percentage of Americans working since 1978.

Instinctively the American people understand they are being lied to by their elected officials when these officials protest they are working to right our fiscal predicament. The people may not know exactly what procedural game is being played, but they understand we don't pile up an $18 trillion debt without a whole lot of bipartisan cooperation and collusion.

Some months after the debt ceiling debate, I attended a fundraising dinner in California with a self-described moderate Republican donor who was decidedly skeptical of the radical tea partier that he believed me to be.

But as I described what was happening in Washington, and as I related the story of the debt ceiling just as I've done here, I could see a change in his expression. The change became more pronounced as I detailed the duplicity of the Republican establishment.

Over dessert, I described the Republican leadership's argument that we should all affirmatively consent to lowering the threshold for Harry Reid to raise the debt ceiling, because then we could all vote no and tell our constituents we actually opposed *what we had just consented to allow happen.*

Quietly, almost under his breath, he whispered, "The bastards." He repeated those words three times.

———

I didn't go to Washington to join a club. Though I try to treat every senator with civility and respect, I didn't ever intend to become part of what Reagan once referred to as the "fraternal order" of Republicans. That is certainly not why I leave my wife and two little girls in Texas every Monday morning—because I prefer the warm embrace of Washington.

Our country is in crisis. We cannot keep doing what we're doing. There comes a point where the hole is too deep, where the debt is too much, where our liberties are too far gone. I don't believe we're there yet, but we're close. It is now or never.

In the Senate, I've tried to do two things: tell the truth, and do what I said I would do. We should expect that from every single elected official.

Today is a time for truth.

That is why I am writing this book. Because public officials need to start acting with the best interests of their constituents, not just the next election, in mind. Because we need to hold them accountable and force them to be honest with the American people. The only way to fix Washington is to shift decision making from the smoke-filled rooms in Washington, D.C., back to the grassroots and the people.

The lobbyists and politicians will fight back. They like the status quo, and they want to keep growing Washington and maintaining their power. But they can be overcome. The American people, if informed and if motivated, can be empowered. They can defeat Washington's corruption.

I'm blessed with a profound appreciation for how great this country is, and about its potential, because of the unusual path I traveled to come to Washington in the first place. It's a story that started long before I was born. As we start this book, I want to share it with you.

Along the way, I also want to highlight stories of people who have inspired me—truth tellers who challenged the conventions of their day, sometimes at great risk. I would not suggest comparing my short tenure in the Senate to the heroic battles of these titans. But these are men and women whom I respect and admire. All of them shared a deep conviction that if you tell people the truth, the people will find their way to the right decisions.

A Time for Truth

CHAPTER 1

The Beacon

Guillermo Fariñas was a committed communist and devotee of the charismatic leader of pro-Marxist rebels in Angola—a man named Fidel Castro. Fariñas believed in Castro because he believed in communism. He trusted in its promise to end poverty and fell for its critique of capitalism. He mistook its propaganda for news and its predictions for promises. His problem wasn't that he didn't care about the truth. His problem was that he didn't know he hadn't found it.

And in the late 1970s he was still a boy, not yet eighteen.

While in Angola, Fariñas—who had been dubbed "El Coco" because his large, bald head looked like a coconut—was wounded on the front lines. Because he had proven himself on the battlefield as one of the most committed to the communist cause, a new global order, he received medals from Castro's army. Then its leaders sent El Coco to the Soviet Union for special military training.

El Coco planned to be a leader of a worldwide movement, a new socialist collective that people like Castro told him would help impoverished people all over the world. But then the young man came across something he did not expect. Something that affected him profoundly.

It was a book by a British author who had been banned in much of the communist world—and is still banned in many totalitarian regimes today.

The book was by George Orwell. And it was called *Animal Farm*. Written by an Englishman during his nation's wartime alliance with the Soviet Union, the book used farm characters—a ruthless pig named Napoleon and a courageous one who challenges him, named Snowball—to display the ruthless and perverse nature of Joseph Stalin's communist totalitarianism. The book demonstrated the perils of forced equality, the power of false propaganda, and the toll that iron-fisted rule exacts on individual freedom. The classic work, which ends in Napoleon's tragic triumph, struck a chord deep within the young El Coco. Slowly he came to realize he was on the wrong side. And he began a new life that would be guided by one simple rule: Seek the truth, and tell the truth, no matter the consequences.

For decades, Cuba's secret police had tried to carry out its program of systematic terror in, well, secret, and El Coco did everything possible to change that. He understood that regimes that rely on secrecy fear exposure and attention as much as they fear any weapons, and he knew that hunger strikes offered him the best way to shine a light of the truth of what life in Cuba was—and is—really like. After El Coco returned home to Cuba, he was confronted by a real-life Animal Farm. He watched the brutality of Castro's security forces and suffered periodic imprisonment for daring to offer opposing political views. And he began a series of grueling hunger strikes designed to highlight the abuse, detention, and murder of his fellow dissidents on the island.

In 2010, the European Parliament awarded Fariñas the Sakharov Prize after an extended hunger strike embarrassed Fidel and Raul

Castro into releasing fifty-two political prisoners. Fariñas, however, cautioned that while he was happy they had been released, until there was real political reform in Cuba such episodes were largely cosmetic and would be exploited by the regime: "For the government," he told the BBC, "political prisoners are a bargaining chip with the civilized world. We are slaves that they can sell when they want."[1]

In the course of his journey to receive the Sakharov Prize—a journey that Cuba had forced him to delay for three years—El Coco visited Capitol Hill. Some people probably thought his presence in America was evidence of a new softening of Cuba's hard-line stance, and an indication that this last front of the Cold War was finally breaking down. But when I met with him and his fellow Cuban dissident Elizardo Sanchez in my Senate office, they assured me that such wishful thinking was far from the truth.

Don't be fooled, they told me flatly. Their trip was yet another example of the Castros' exploiting dissidents to trick the United States into thinking they had reformed, while in reality they were consolidating their power. Cuba was "a big jail." The Castros' control over it was pervasive. Men and women on the island struggled every day with choices that balanced their dignity with their safety from the secret police. Some choose to become part of the system; others choose to feign mental illness; still others, like El Coco, choose to tell the truth about Cuba and join the opposition, even though it often means abandoning their homes and becoming the targets of relentless mental and physical abuse.

Fariñas and Sanchez warned that the Castros were looking to the example of Vladimir Putin in Russia, who had also projected the illusion of change to the West and won a number of economic and security concessions—and all the while had been carefully orchestrating the reimposition of a centralized control not seen since the days of the Soviets. "Putinismo" they called it. Like Putin, the Castros were sure they could exploit our naïveté and extract concessions from America.

A few weeks later, Fariñas and Sanchez were back in Cuba, once again telling the truth, and once again subjecting themselves to harassment and abuse. The wisdom of their predictions about the Castros' strategy was vindicated by the president's stunning announcement on December 17, 2014, that the United States would unilaterally relax the economic embargo on Cuba in exchange for yet another artificial round of political prisoner releases. No meaningful reforms were demanded, and no guarantees were made to prevent Cuba from arbitrarily detaining the released prisoners—and their allies—in the future. And now America would help the Castros pay for their imprisonment. Just as El Coco predicted.

———

Some years back, my wife, Heidi, and I were having dinner with friends. During the course of our conversation, a heated argument on some topic or another, one of them stopped me. "Ted," he asked, "when did you first get interested in politics?"

I replied, "I'm not sure. To be honest, I don't ever remember a time when I wasn't."

This was not the politic answer. Anyone considering running for office, as I was at the time, is supposed to act totally disinterested in the political process, to pose as the reluctant public servant only answering the call because the people need him or her so desperately. But that wasn't the truth. Not for me.

I told our friends that I'd been interested in politics since I was a young child. "I really don't know why that is," I added.

At this point, Heidi flashed the look that only a spouse can give.

That of the wife who sees things that are blindingly obvious, except that we are too obtuse to see it ourselves.

"Ted, it's really no wonder," she said, laughing. "Think of the family you were raised in."

I guess most people think of their family as pretty typical, and I was no exception. But not every family is populated by people like Rafael Cruz.

My dad, a Cuban immigrant who sometimes seems larger than life, has always been my hero. He has always felt a visceral urgency about politics. Having the right people in office was vitally important to my dad, as if it were a matter of life and death. Because for him, in a very literal sense, it was.

There isn't a day that goes by when my thoughts don't turn to a boy with jet-black hair, a curious mind, and an instinct for rebellion who was just emerging into manhood. He was born to a middle-class family in Cuba and had earned straight A's in school. His future was filled with possibility, and he might well have prospered under the regime of Fulgencio Batista.

But he and his friends quickly realized the cruelty of Batista's totalitarianism. He watched in horror as military police beat the government's opponents. Along with other young students, he secretly allied with an underground movement to replace a cruel and oppressive dictator. The movement was led by Fidel Castro, whose own capacity for tyranny and terror was not yet known—at the time he seemed to hold the promise of freedom. That dark-haired boy became a guerrilla, throwing Molotov cocktails at the buildings of Batista's regime, whatever the resistance needed.

As the Cuban army began a crackdown, a number of the young man's cohorts were arrested. Some were shot dead in the streets—a fate he often expected to share. "Praise God that He had a better plan," he sometimes tells people. The young man was Rafael Bienvenido Cruz. My father.

———

As it was for so many who hailed from the 780-mile-long strip of mountains in the Caribbean, the island of Cuba began as a place of refuge for the Cruz family. It eventually became a nightmarish horror, a place where possibilities are upended and where hope is routinely and systematically destroyed by an all-controlling and corrupt central government.

Discovered by Christopher Columbus in 1492, Cuba became one of the earliest Spanish colonies, and from there the Conquistadors launched expeditions into Mexico and Central and South America. These ruthless men slaughtered much of the native Cuban population. In fact, the name of the city where my father was born, Matanzas, literally means "massacre." The Spaniards killed virtually everyone in the original native village.

For centuries Cubans bristled under the Spanish Empire. Toward the end of the nineteenth century, revolution was brewing. From his exile in the United States, Cuban poet José Martí led impassioned efforts to seek support for a war of independence to free Cuba from Spain. The Spanish government had made it illegal for Cubans to own weapons; only Spanish soldiers were allowed to own guns. And so the Cuban rebels improvised. They used what they had—machetes, the two-foot-long knives that they used to harvest sugarcane. At the cry of *"al machete,"* the rebels would storm the Spanish army against their musket fire.

When the USS *Maine* was sunk in the port of Havana in 1898, it precipitated the Spanish-American War. At President William McKinley's urging, Congress declared war against Spain in April of that year. American troops landed near Santiago, on the southeastern end of the island, and made Guantánamo Bay their base of operations (that base remains a renowned U.S. military outpost to this day). During this short-lived war, Teddy Roosevelt achieved national acclaim by leading his "Rough Riders" to successfully conquer the Spaniards and take San Juan Hill. A few months later the United States and Spain signed a treaty granting Cuba its independence.

It was to this fledgling nation that my great-grandparents arrived in 1902 from the Canary Islands. Agustin and Maria Cruz boarded a ship with their infant son, Rafael, bound for the New World. Their hearts filled with anticipation, a sense of adventure, and a single ambition: to own their own farm. After a decade of hard labor, Agustin had saved enough to buy a few acres of land.

But Agustin died in 1917, one of the millions of lives claimed by a worldwide influenza epidemic. Maria was left a widow with six children. Overcome by despair and depression, Maria ended up being swindled out of their farm. The family had lost everything. Maria and her children were forced to move to a sugarcane plantation. In exchange for her older boys, including Rafael, cutting sugarcane all day, they could live in a hut with a dirt floor.

Life at the plantation was hard. There were a hundred or so huts built in a circle, forming a small village. There was only one general store in the village, owned by the sugar mill, where the workers bought everything from food to tools to clothing and shoes. The store gave the families credit, and the sugar mill paid their salaries through the general store, which then took the money to pay their debt and (in theory) give them any remaining money. But, of course, no money ever remained, and the arrangement essentially led to perpetual servitude.

One day, a bus driver came by the plantation trying to round up people to attend a political rally in the city of Matanzas. The driver offered anyone who came with him five dollars and a sandwich. To my grandfather, that sounded pretty good. He got on that bus, never to return to the plantation. My grandfather managed to get a job working at a fruit stand on the beach near Matanzas. Since he was broke, he was grateful that the owner let him sleep on the floor of the fruit stand.

He sent some of his salary back to his mother on the plantation, and in time, my grandfather was able to save a little money for himself as well. The owner of the fruit stand expanded, opening a restaurant across the street. And my grandfather took his meager savings and bought the fruit stand. Rafael Cruz became a small business owner . . . of a magnificent beachfront fruit stand.

Rafael soon married and had a son, but his first marriage didn't last. He then met Laudelina Diaz, my grandmother. She lived with her parents, Juan and Lola Diaz, and her seven brothers and sisters, on a small ranch about a quarter mile down the road. She was eleven years younger than he and was a schoolteacher. Laudelina was blessed with a nearly photographic memory. She loved her sixth-grade students, and they felt the same way about her. Indeed, decades later, when she was retired and living in Texas, she would repeatedly be surprised to see former students from Cuba who had tracked her down to thank her for the loving care and inspiration she had given them. For five years, Rafael courted Laudelina, and then they wed.

Although he had only a third-grade education, Rafael's business prospered at first, and he was able to grow the fruit stand into a proper grocery store. He decided to bring in his youngest brother to join him in running the store, but the young lad turned out to be more interested in women and liquor than in being a responsible business partner. Nor were matters helped by Rafael's generosity. When a strike at a nearby textile factory left a substantial number of workers without income, Rafael offered credit at his store to all the factory workers. But few felt honor-bound to pay him back when the strike was over. The combination of his irresponsible brother and the mounting delinquent accounts forced Rafael to close the grocery store.

Needing to find a new way to provide for his young family, Rafael became a commission salesman for RCA. One of his favorite sales methods was something called the "puppy dog close" (derived from the easiest way to sell a dog—letting a family simply try the puppy for a few days). When television came into Cuba, my grandfather would urge prospective customers to try out a TV for a little while, for free. At any point in time he had a couple of dozen TVs on loan to different people. He would run across you on the street and say, "There is a great boxing match on TV this Saturday," or "Elvis Presley is going to be on a TV show next week," followed by "Why don't I lend you a TV so you and your family can watch it?" They would often reply, "But I don't want to buy a TV," to which my grandfather would respond, "Who is talking about buying? I have a half a dozen TVs and you can keep one for a couple of weeks. I'll pick it up when I need it." Then, after a couple of weeks, he would come by their house on Saturday morning . . . right in the middle of Saturday cartoons. He'd knock on the door, walk into the living room, and reach over and unplug the TV. Then he'd pick it up and begin walking out of the house.

Inevitably, the kids who had been watching cartoons would scream and cry, and the parents would be desperate to mollify their children. My grandfather sold a lot of TVs on Saturday mornings.

On March 22, 1939, Laudelina gave birth to her first child, Rafael Bienvenido Cruz. *Bienvenido* is Spanish for "welcome," and true to his name, Dad was very much a welcome addition to the young couple's home. A few years later, she had a second child, Sonia Lourdes. Sonia is a firebrand—my "Tía Loca"—whom I love and admire.

Growing up in Cuba in the 1940s could be challenging, but it was also wonderful. My dad loves the sea, and he spent count-less hours at the beach, fishing with his father and a 44-pound-test monofilament line that they would hold directly in their hands. Even a small fish would cause the line to burn and cut your fingers as you let it slide between your finger and thumb while working the fish. My grandfather's and father's fish stories more often than not included showing the cuts on their fingers to demonstrate the ferocity of Cuban fish.

One time, my father and a friend decided to go fishing for a shark. The two boys got in a wooden rowboat, brought some rope with a chain leader, put a hunk of tuna on a large hook, and tossed it into the bright blue water. Sure enough, they ended up catching a shark, about six feet long. The shark was none too happy, and he charged their boat (itself only about ten feet long). The shark's head rammed the boat and broke a hole its bottom, which resulted in water flooding the boat. The boys panicked. They cut the rope, and one boy began bailing water while the other rowed as fast as he could.

The boat sank a few hundred feet from shore, and the boys swam and towed the boat to the beach, terrified that an angry shark was behind them looking for retribution. Thankfully, the shark had other priorities, and they emerged unscathed.

But life in Cuba during those years wasn't just about fishing ad-ventures. Revolution was in the air.

Fulgencio Batista, a former Cuban army sergeant, led his first military coup in 1933, when he overthrew the democratically elected president of Cuba. Exiled in 1944, he returned and staged another coup in 1952. Firmly ensconced in power, Batista became a brutal dictator. With the military as his enforcers, Batista behaved like a mafioso, extorting protection money from local businesses. He showered privileges on a crony class in Havana and imprisoned and tortured those who crossed him. And the mafia comparison was not just figurative: Batista embraced American mobsters from Chicago and Las Vegas, and soon prostitution, drug traffic, and casinos proliferated.

In the *Godfather* movies, Michael Corleone flourished in Cuba; in real life, Batista used his mob profits to enrich himself and to buy a formidable arsenal, which he used to instill fear and impose order. Oppression breeds resistance, and the Cuban rebels were led by a young revolutionary named Fidel Castro.

Castro first burst onto the scene on July 26, 1953, when, as a young lawyer, he led a small band of armed men in an attack on the Moncada army garrison in the city of Santiago, on the southeastern coast of Cuba. Most of those men were killed, but Castro and a few others were captured and imprisoned. After being convicted, he was sent into exile in Mexico.

In protest, student demonstrations erupted all over the island. Batista's army broke up the demonstrations by beating the students with billy clubs.

In the United States, student councils typically focus on student life, planning proms, and promoting school spirit; that certainly was our emphasis when I was on the student council in high school. It was not exactly the province of the "cool kids," and few American student council leaders would be confused with revolutionaries.

In Cuba, the demonstrations—and ultimately the revolution— were in many ways led by high school and college student councils. At first the demonstrations were uncoordinated, but as the abuses of the Batista regime increased, they became larger and better organized. Eventually the opposition developed into a full-fledged underground movement, carrying out sabotage, propaganda, and attacks on political leaders.

At the time, none of the young men who joined the resistance knew that Castro was a communist. Castro was seen as a freedom fighter, an inspiring figure to Cuba's restive youth, or, as my Dad puts it today, "fourteen-year-old boys who didn't know any better."

———

The anti-Batista underground consisted of a series of units operating in a semi-independent manner from one another. Secrecy was paramount, and after an initial period of training, my father—a high school student council leader—began recruiting others and forming his first unit. For reasons of security, the members of his unit did not know to whom my father reported. Their unit concentrated on propaganda, acquisition and movement of weapons, and acts of sabotage.

In September 1956, my dad enrolled at the University of Santiago and reported to Frank Pais, the provincial leader of the 26th of July Movement (named for the date of Castro's first failed attack). Soon they began preparing for Fidel Castro's imminent landing in Cuba with a group of eighty-two rebels; Castro and his group were traveling from Mexico on a sixty-foot cabin cruiser, the *Granma*. On the evening of November 29, Pais and his men were given final instructions and divided into two groups. The group my father was in was to join Castro and his rebels as he again stormed the Moncada army garrison the next morning. Frank Pais led the second group, which was to attack the police headquarters in Santiago at the same time.

As morning arrived, word came to my father's group that for some reason the landing had not occurred and they were to scatter as quickly as possible. Later, they would learn that engine trouble had delayed the *Granma*. Their group dispersed. But the other group, apparently, did not get word.

The second group of rebels instead carried out their planned attack on the police headquarters, where they encountered fierce resistance. All of the students were killed, as was their leader, Frank Pais.

Two days later, on December 2, Castro and the *Granma* landed in Cuba. But by that time the army was on high alert and Castro had no choice but to head for the mountains. Of the eighty-two revolutionaries traveling with Castro, only nineteen reached the sierra. The rest were either killed or captured.

With the army rounding up rebels everywhere, my father and three of his friends decided to flee Santiago by car. On their way out of the city, however, they were stopped by an army patrol and captured. The soldiers took them to the Moncada garrison. Finally and ironically they would make it into the garrison they had planned to storm.

As they were taken off the army truck, my father could hear someone say under his breath, "What a shame, they look so young." As my father and his men entered the garrison, soldiers jeered and yelled at them. "*¡Al paredón! ¡Al paredón!*" (To the firing squad, to the firing squad.)

My father thought that was the end. He resigned himself to death. "After all," he thought to himself, "we're young. We don't have a family. No one depends on us. What does it matter if we lose our lives for freedom?"

But, thankfully, the Lord had a different plan for him. As they were marched into Moncada army garrison, they spotted a fellow student from the university who was the son of an army major. Together they cried out his name, "Erasmo, Erasmo!" It turned out he had been picked up by mistake and was just being released.

Erasmo walked over and asked what was happening. My father and his friends said they didn't know why the army had taken them. They asked for his help. Erasmo had no idea that they were in the underground. He vouched for all four and they were released.

Were it not for that chance encounter, that providential intervention, my father likely would have been executed that afternoon. Instead, he returned to Matanzas and resumed control of his rebel unit. He formed a second one, focused on sabotage throughout the province, especially trying to disrupt communications and transportation.

For the young revolutionary, one of the greatest challenges was knowing whom to trust. Recruit the wrong person, trust the wrong person, and the results were fatal.

He worked hard to recruit a young man who seemed eager to join the revolution. Too eager, it turned out. The man was a government informant.

By the time he was seventeen, my dad had been fighting for several years. Then, one day, he disappeared.

My grandfather knew that his son was in the underground. And he knew that his disappearance was a very bad thing. So he began to search for him.

He went from jail to jail, eventually finding my father in an army garrison in Matanzas. It wasn't pretty. The informant had turned him in. They threw him in a rotten cell, acrid with the smell of blood, grime, and urine. Men with clubs beat him. His captors broke his nose when they kicked him in the head with their army boots. They bashed in his front teeth until they dangled from his mouth. In each round of beatings, the pain was unbearable. Then it subsided altogether because the pain had completely numbed him. He often couldn't feel his hands or his legs.

My father rarely talked about what happened to him, but years later, when I was a teenager, he and I went to see the movie *Rambo*. There's a harrowing scene in that movie where Rambo is tied to some metal bedsprings and his Vietnamese captors torture him with electricity. That night, my parents had some friends over for dinner, and I remember my dad talking about the movie he and I had seen that afternoon. He said, "You know, the Cubans weren't nearly that fancy in their torture methods. They would just come into your jail cell every couple of hours and beat the crap out of you. And then they'd do it again and again."

Throughout the ordeal, my dad refused to tell his interrogators who else was in the underground. He had heard of too many rebels who had broken and had been shot, with their bodies dumped in the street. That fear kept him from breaking.

He was then dragged into the office of a colonel.

"We're letting you go," he said to Dad. "But if another bomb explodes in this city, we're coming to get you."

"Well, how can I be responsible for what other people do?" my father asked.

"I don't care," the colonel said. "I'm coming to get you."

My dad was released so they could follow him. He was put under constant surveillance in the hope that he would lead them to other revolutionaries.

When he came home, my grandfather was adamant. "They know who you are now," he told his son, with fear in his eyes. "It's only a matter of time before they kill you. You've got to get out."

Shortly thereafter, a woman from the Castro underground visited his house surreptitiously. She said he was under twenty-four-hour surveillance, the army was watching him, and it was too dangerous for other rebels to be seen with him. My dad asked if he could join Castro in the mountains and keep fighting, but he was told there was no way to get to the rebels.

And so my dad decided to flee Cuba. He had been a good student in high school, graduating first in his class. So in 1957, he applied for admission to three American universities: the University of Miami, Louisiana State University, and the University of Texas. Texas was the first to let him in, which set our family's roots in the vast and opportunity-rich Lone Star State. With the letter of acceptance from Texas in hand, he went to the U.S. embassy and received a student visa. All he needed now was an exit permit from the Cuban government. That was not easy to get, especially for a young man who had been arrested as a rebel. Fortunately, the Batista regime was nothing if not corrupt. A lawyer friend of the family quietly bribed a Cuban government official, who stamped my father's passport to let him out.

Early one morning in August 1957, my dad lay on the backseat of his father's car inside their garage. My grandfather pulled the car out of the garage and drove to Havana, to the docks in the harbor. As my father prepared to leave, he was scared. His mother and sister cried and embraced him. He was leaving behind everything and everyone he knew. But there was only one place my father wanted to go.

As the ferryboat traversed the ninety miles between Cuba and America, my father pondered the life he had left behind. With the smell of salt water in the air, the smell he had grown up with every day, he thought of his friends, his parents, and his kid sister. He didn't know if he would ever see any of them again.

Across those choppy seas, where so many Cubans have lost their lives trying to escape to freedom, powerful emotions gripped my dad. He felt profound guilt at having left the revolution, having left his comrades in arms. He felt sadness and uncertainty as to what was to befall his beloved Cuba. He was afraid, deeply afraid, as to how he would survive in a new country where he was almost totally alone. And yet he was excited. He was headed to America.

It is difficult for many of us to fully comprehend what a beacon of hope this country offers the rest of the world. There is no other place on earth that would have welcomed so freely to its shores a man like Rafael Cruz. He was eighteen, penniless, and spoke no English. He owned three things: the suit on his back, a slide rule in his pocket, and a hundred dollars that my grandmother had sewn into his underwear.

America, quite simply, saved my father. America gave him a chance.

After his ferry arrived in Key West, my dad then boarded a Greyhound bus for Austin, Texas.

He found a place to live, and got a job washing dishes in an Austin diner called the Toddle House. The job paid fifty cents an hour, but it was perfect because it didn't require him to speak English. The best part about working at the restaurant, however, was that employees were allowed to eat for free on the job, which was a pretty good deal for an impoverished student. Indeed, he rarely ate anything unless he was on the job. Since he liked to eat seven days a week, he began working seven days a week as well.

At the same time, he went to school full-time. That presented an urgent challenge: learning English. All of his classes were in English, and he didn't know the language. And if he flunked out of the university, he'd lose his student visa and have to go back to Cuba.

So he signed up for Spanish 101. While the professor was teaching Spanish ("*leche* means milk"), my dad did his best to reverse-engineer the class ("*milk* means leche"). To boost his grades, he took lots of math classes, since those were less language dependent. And he spent what little money he had going to the movies—a lot. Every Saturday he'd stay in the theater and watch the same movie three times in a row. The human brain is remarkable. By the third time, simply from context and intonation, he'd begin understanding parts of what was going on. And, quite quickly, he began learning English.

After my father learned English, he began going around to

Rotary Clubs in Austin, speaking about the Cuban revolution. My father is a passionate and powerful speaker, and even in broken English, he would urge Austin businesspeople to oppose Batista and support the revolution and its leader, Fidel Castro.

In 1959, while he was in Austin, the revolution succeeded. Castro seized power and declared to the world that he was a communist. That summer, my dad returned to Cuba thinking he was finally free to visit his family, but he was horrified by what he saw. It quickly became evident that Castro was even worse than Batista had been. My father's former hero began seizing people's land and publicly executing dissidents. Nothing could be further from my father's ideals than the tyranny of communism.

Indeed, as Castro's brutality became apparent, my dad's kid sister Sonia followed her brother's path. But she became part of the counterrevolutionary struggle. She fought against Castro, trying to topple him from power. She engaged in sabotage, lighting the lucrative sugarcane fields on fire. Sonia and the rebels fled to the hills, living for months in caves and under the stars

And, sadly, she too faced prison and torture. Castro's goons threw Sonia and her two best friends in prison—all of them were teenage girls—and brutalized them. I love my Tía Sonia—she's a wonderful, passionate, loving person—and we don't talk about what she experienced in that Cuban jail.*

Once Castro became dictator, he put in place an Orwellian regime of systematic oppression. For the state to be supreme, all other loyalties had to be destroyed—from family to religion. Children were taught to spy on their parents and to report anything they said that was disloyal to Castro. Churches were closed, and faith was

---

*In 1962, with a forged passport to escape Cuba, she fled to America to join her brother. My grandparents joined them eight years later.

aggressively persecuted. My grandmother told a story of Castro's soldiers coming into classrooms carrying machine guns. They told the kids to close their eyes and pray to Jesus for some candy. The children did so, and no candy appeared. Then they told the kids to pray to Castro for candy; while their eyes were closed, the soldiers quietly put candy on each child's desk.

Schoolteachers were required to indoctrinate their students with communism. But my grandmother wanted nothing to do with poisoning children's minds, and so she resorted to an extreme measure: She feigned insanity. She began screaming gibberish and foaming at the mouth, running out of her classroom. She enlisted a doctor friend to give her a bogus diagnosis of insanity to get her out of teaching school. It always amazed me that my grandmother was willing to endure the stigma and ridicule of being thought crazy rather than be complicit in corrupting young children.

Back in Austin, my father sat down and made a list of every place he had spoken in support of Castro's revolution. He made a point of going back to each and every one of them, standing before them, and apologizing. He would say, "I misled you. I didn't do so knowingly, but I did so nonetheless. And for that, I am sorry."

I always admired that about my father. Many people were horribly wrong about Fidel Castro, and many still are today. This is especially true among the fashionable circle of liberals who have decided to make Castro a sort of socialist folk hero. Leftist members of Congress have toured Cuba and come back raving about Castro and the wonders of his communist system. And of course Hollywood celebrities who have based their political philosophy on the idea that the American government is the world's villain chime in with this chorus.

During the Cold War, there was a term for people who were manipulated and duped by a dictatorship so they could spread propaganda about it to the rest of the world—"useful idiots." In Cuba,

Americans welcomed into the country are carefully monitored by the regime. A taxi driver, for example, might pick someone up at the airport in Havana and offer to take them anywhere they want to go. If the visitor says, "I want to see a typical Cuban farm," the driver might reply, "Sure. I will take you to one." The taxi will go to a beautiful farm filled with livestock, food, and happy Cubans, all of whom extol the Castro government and express sadness that Americans have such a false view of the communist system. What many visitors don't realize is that this taxi driver was not some random Cuban and this farm was not picked by chance. Everyone they are meeting is an agent of the state. The same is true about hospitals that filmmakers are allowed to "spontaneously" visit; they look top of the line and the quality of care seems excellent. But this too is a ruse. The state-run health care is a two-tiered system: one for corrupt party officials, for whom no expense is spared, and the other for everyone else. Nothing happens in Cuba, as in any dictatorship, without the close, watchful eye of the state, even if that eye is hidden to oblivious visitors who want to believe Castro is a hero and America is a bully.

My father was also fooled by Castro, but only for a time, and only in his youth. Few people would go out of their way to atone for such a mistake, but he did, by denouncing his own public statements and asking for forgiveness. That must have been difficult, and it's one of the many reasons I've always admired my dad's character.

The importance of speaking the truth was one lesson my father learned from Castro's rise to power; another was what a society that steadily erodes its citizens' individual freedoms looks like. He's seen it happen before.

Barack Obama, noting his own rise from humble beginnings, has observed that "in no other country on earth is my story even possible." My family can relate to that sentiment. In no other country would Rafael Cruz's story even be possible.

And yet, in many ways, his story is commonplace. All of us are the children of those who risked everything for liberty. It's what ties Americans together. What has always been special about America—what has always been part of our national DNA—is a profound love of liberty and opportunity, an embrace of the unlimited potential of free men and free women.

But, as Reagan powerfully observed, "freedom is never more than one generation away from extinction. We didn't pass it to our children in the bloodstream. It must be fought for, protected, and handed on for them to do the same, or one day we will spend our sunset years telling our children and our children's children what it was once like in the United States when men were free."

My dad is particularly emotional whenever this topic surfaces. The freedom of America was the dream that allowed him to endure the brutality of Cuba. It was and is a beacon of hope for all those who, like him, have endured oppression.

"We have to turn America around," he says, his voice quivering. Thinking of the so many others who look to us for inspiration and hope, he continues, "When we faced oppression in Cuba, I had a place to flee to. If we lose our freedom here, where do we go?"

———————— ★ ————————

CHAPTER 2

To the Lone Star State

Nearing seventy, the tall and imposing Texan was more weary and embattled than in the days of his splendid youth, but he still managed to carry himself with the dignity and pride that had sustained him as the founding father of the state he so cherished.

Sam Houston, elected to the governor's chair by 56 percent of the people a little more than a year earlier, arrived at his office early. As was his custom, he was carrying his lunch basket, so he could work from his desk. There was little time for idleness. Texas was under siege, and as he told his political opponents, Houston intended to work for its salvation just as long as he was able.

Upon entering his office, however, he realized at once that the threats made against him hadn't been idle. Another man was seated at his desk, in his chair. This was not Sam Houston's office anymore.

The former president of the former Republic of Texas, the state's first U.S. senator once it joined the Union, the once-popular governor of the Lone Star State, was now unemployed. Fired, to be exact. Thrown out of office by a Confederate convention that he refused to support or legitimize, whose power to speak for the people of

Texas he would vociferously deny. Now Ed Clark, the Confederates' appointed governor of Texas, and usurper of his office, was sitting in his chair.

"Well, Governor," Houston said, his voice filled with sarcasm as he uttered the title, "I hope you will find it an easier chair than I have found it." [1]

"I'll endeavor to make it so," Clark replied, "by conforming to the clearly expressed will of the people of Texas."

The words cut into the proud Houston, who had vigorously opposed Texas's decision to join the Confederate States of America in February 1861. For Houston, a slave owner who nonetheless bore no love for the abhorrent institution and opposed its expansion, the American republic mattered more than any grievance nursed by his southern neighbors.

"Upward of forty-seven years ago, I enlisted, a mere boy, to sustain the national flag and in defense of a harassed frontier, now the abode of a dense civilization," he had said in a speech a few months earlier. "When, in 1836, I volunteered to aid in transplanting American liberty to this soil, it was with the belief that the Constitution and the Union were to be perpetual blessings on the human race—that the success of the experiment of our fathers was beyond dispute, and that whether under the banner of the Lone Star or the many-starred banner of the Union, I could point to the land of Washington, Jefferson, and Jackson, as the land blest beyond all other lands, where freedom would be eternal and Union unbroken." He had implored Texans not to trade the Union "for all the hazards, anarchy and carnage" of civil war—"a leap in the dark—a leap into an abyss, whose horrors would even fright the mad spirits of disunion who tempt you on."

It was to no avail. For Confederates, the final straw had come when Houston refused to swear an oath to the Confederacy—the only southern governor to so refuse. There were many other Texans who did not share the Confederates' views about the war, or about their leader. Some urged Houston to fight on: "Save Texas for us if you can."

But he knew the cause was lost. Too many Texans had already decided to take that plunge into chaos. Houston could not bear to watch with his own eyes what would happen to his beloved Texas. So he and his wife decided to leave the state they loved.

All along his route to Alabama, the deposed governor was greeted by his fellow Texans. Some made death threats; others hailed his name, and urged him to speak.

In April 1861, as the bombardment of South Carolina's Fort Sumter began and the war Houston long had feared got under way, he found himself back in Galveston, Texas. Word of his arrival spread, and a large and angry crowd of Confederates, believing Houston a traitor, gathered outside the balcony of his hotel. In fear of Houston's life, friends urged Houston not to speak to them. But Houston was never one to run from conflict, or hide from the truth.

Instead, he offered a famous warning. "Some of you laugh to scorn the idea of bloodshed as the result of secession," he said. "But let me tell you what is coming. Your fathers and husbands, your sons and brothers, will be herded at the point of a bayonet. After the sacrifice of countless millions of treasure and hundreds of thousands of lives you may win southern independence, but I doubt it. The North is determined to preserve this Union. They are not a fiery, impulsive people as you are, for they live in colder climates. But when they begin to move in a given direction, they move with the steady momentum and perseverance of a mighty avalanche."

His speech was met with jeers and catcalls. Sam Houston had helped build the state of Texas, and now he was branded a traitor.

He would die two years later, in July 1863, penniless and thought to be disgraced, unable to bear witness as his warnings of disaster for the Confederacy proved themselves prescient. Only a handful of Texans attended his funeral. His tombstone reads: "A Brave Soldier. A Fearless Statesman. A Great Orator—A Pure Patriot. A Faithful Friend, A Loyal Citizen. A Devoted Husband and Father. A Consistent Christian—An Honest Man."

———

If my dad has been a hero to me, my mother has been a best friend for as long as I can remember. Eleanor Darragh is a woman of unconditional love and deep compassion, but also of formidable intellect and unshakable strength. These qualities she often deployed against some pretty tough odds.

Her parents were Irish and Italian. In the late 1800s, my grandmother's father, Dominic Ciccini, came to America from Naples, Italy, as a teenager. In the melting pot of nineteenth-century America Italian and Irish immigrants often intertwined, and so it was with my family. Dominic fell in love with and married Mary Lunergen, who had been a teenage immigrant from County Tipperary in Ireland. After a few years, for reasons lost in family lore, Dominic changed their family name from Ciccini to the more Americanized Cekine, pronounced "See-kine."

Dominic and Mary began having children, and on April 10, 1912, my grandmother, Elizabeth Eleanor Cekine, was born in Wilmington, Delaware. By the time Elizabeth arrived, the house was full of kids—she was the second youngest of seventeen children (ten of who survived to adulthood). It was your typical Irish-Italian family, loud, boisterous, working class. Indeed, we sometimes joked that there were only two families in Delaware: the DuPonts, who owned the factories, and my family, who worked in them.

Not everyone in Mom's family worked a blue-collar job; some wound up on a shadier side of the law. My mother's uncle John and uncle Albert both ran numbers, with Albert heading up the numbers racket in Wilmington. And Mary, the matriarch, developed a surefire way of carrying the numbers without being stopped by the police: She would conceal them between two soup pots, nested together, as she walked down the street.

At the age of nineteen, young Elizabeth married Edward John Darragh, also of Irish descent.

Edward's parents were from Chester, Pennsylvania, and his father had enjoyed some modest success as a small business owner, running a brewery and bar. But, when Edward was twelve, his father died unexpectedly, and so he dropped out of school after the eighth grade. Thus, at the age of twelve, my grandfather went to work full-time as a clerk with the Reading Railroad (yes, the same one that's on the Monopoly board).

Together, Edward and Elizabeth had three children. The eldest, my mom, was born in 1934. My mother, Eleanor Elizabeth, is pretty, five foot two with brown hair, fair skin, and dark brown eyes. Shy and unusually bright, she excelled in school—something my grandfather thought superfluous for a girl in the 1940s.

Her father was not an easy man. Edward was gruff, opinionated, and often selfish. He drank too much. (Years later, Eddie, my uncle, was such an extreme alcoholic that he lived with his parents his entire adult life, could not work, and died from the ravages of his drinking.) And when my grandfather was drunk, he was a mean drunk. As the eldest, my mom bore the brunt of that.

When my mother was in high school, my grandfather went to work as a traffic manager in Baton Rouge, Louisiana, where my mom went to a Catholic all-girls' school, St. Joseph's Academy. She worked hard and graduated near the top of her class. A few years later, my grandfather was transferred to Houston, and after high school graduation my mother joined her parents there.

My mom decided to apply to the Rice Institute (today it's Rice University, one of the top liberal arts colleges in the country), which was all of two miles down the road. Much to the astonishment of my grandfather, she got in. Nobody in her family had ever gone to college, and my grandfather certainly didn't think his daughter should be the first. His view—which he would express in a drunken, verbally abusive rage—was that there was no need for women to be educated.

But my mother's shyness can be tempered by her strong will. Her Irish tends to come up whenever she sees injustice. So she battled her father. She was going to go to college, she declared, and there was not a thing he could do to stop her.

If she had needed his financial support, that would have been a problem. But the Rice Institute—founded in 1912 by the munificence of businessman William Marsh Rice—had such a substantial endowment that there was no tuition. Every student was on full scholarship. And so my mom could afford to go.

When she arrived at Rice in the fall of 1952, it was a difficult place. Only half of the student body was expected to graduate; roughly 50 percent of the students would flunk out. Moreover, she made her path even more difficult by choosing to major in math . . . almost unheard-of for women in the 1950s.

During college, she lived with her parents and worked summers at Foley's, a Houston department store.

In 1956, Mom graduated from Rice with a degree in mathematics. She promptly went to work at Shell as a computer programmer—a career that hadn't even existed just a few years before. Her father stubbornly thought she shouldn't be working and living away from home, but she again insisted on going her own way. There were very few computer programmers who were women. The oil and gas industry was (and to some extent still is) dominated by men, as was the computer industry; at Shell, she was at the intersection of both.

One need not be a devotee of *Mad Men* to understand what faced working women in the 1950s. Coming out of college, my mom deliberately didn't learn how to type. She understood that men would stop her in the corridors of the Shell offices and ask her, "Sweetheart, would you type this for me?" With a clear conscience she could answer, "I'd love to help, but I don't know how to type! . . . I guess you're just going to have to use me as a computer programmer instead."

I often reflect on the challenges my mother confronted, especially whenever Democratic political operatives and their media allies spin the notion that Republican officeholders somehow have waged a "war on women." It's a made-up attack, poll-tested to try to scare single women into thinking that politicians want to take away their birth control. The attack is nonsense, designed to distract the voters. The alleged threat simply doesn't exist. I've been around conservatives all my life, and I've never encountered a single person who wanted to ban birth control.*

The real war on women today can be found in the policies that have resulted in 3.7 million more women living in poverty under President Obama, and the median annual wage for women dropping $733 over the past six years. Women like my mom want—and deserve—the opportunity to work and excel in an environment that rewards merit, that pays them what they're worth and allows them

---

*An altogether different issue is whether the federal government can force Americans to *pay for* abortion-inducing drugs for others, as the Obama administration has attempted to do to Christian companies like Hobby Lobby and even to the Catholic nuns in the Little Sisters of the Poor (a charity caring for the poor and elderly). Fortunately, the Supreme Court rejected that attempt, ruling that the federal government cannot force us to violate our religious faith.

to develop their God-given talents and provide for their families. And they cannot do that in a stagnant economy, where small businesses are struggling and opportunity is scarce.

————

In 1956, my mom married her first husband, a mathematician named Alan Wilson. Two years later, they moved to Boston, where she was hired at the Smithsonian Astrophysical Observatory as a programmer, tasked with calculating orbits of the Russian satellite Sputnik. It was pretty heady stuff for an Irish-Italian girl who was the first in her family to go to college.

In 1960, they moved to London, and five years later my mom gave birth to a son, named Michael. For my mother, 1965 was a year of celebration and tragedy. My mom loved Michael and rejoiced in motherhood, but that December she awoke one cold morning to find that he had died in his sleep.

Losing Michael to crib death broke my mother's heart, and had a profound effect on her, so much so that I never even knew that I had had a brother until I was a teenager, when my mom finally told me the story. I have to admit, it was pretty surprising, disconcerting even, to be in high school and discover you had a brother you never knew about. (When I was in elementary school, I repeatedly asked my mom to "please make me a brother"; since she was in her mid-forties, she didn't oblige.)

To this day, my mom is pro-life at a deep, emotional level. Having lost her first child, she understands firsthand that every human life is a precious, unique gift from God. She cannot imagine—it literally makes her tremble—taking the life of an unborn child. What she would give to have Michael back. I so wish my mother didn't have to endure that pain.

And the heartbreak also ended her marriage. In 1966, she and

Alan divorced, and she moved to New Orleans, where she went to work for Geocom, a geophysical data processing company.

And at Geocom, she met a man named Rafael Cruz.

———

In my first conscious memory, I was causing trouble. I was in the grocery story as I put a kazoo to my mouth and blew it, repeatedly, loudly, and to the growing irritation of my mother, for whom an Irish-American temper is not just an ethnic stereotype.

She informed her two-year-old son that if I failed to stop the noise, there would be consequences, namely that she would take me out of the grocery store, drive me home, and give me the spanking I richly deserved. I blew the horn.

My mom stopped shopping, left her cart in the aisle, picked me up, and took me home. In the car, my very first memory is trying my very best to turn the hand of fate. To strike up a conversation with her. To tell a joke. Anything to talk my way out of the punishment. But my mother bested me in determination. She also was a woman of her word. My spanking was forthcoming.

———

I was born on December 22, 1970, in Calgary, Alberta, a growing metropolis carved from the cusp of the Canadian Rockies. After meeting each other at Geocom in New Orleans, my parents had moved to Canada to continue working in the oil and gas industry. They formed their own seismic data processing company to help oil companies search for new reserves. They were both mathematicians and computer programmers, and together they wrote the proprietary software for their business. The name of the company was R. B. Cruz and Associates.

My parents were two strong-willed people struggling to earn

a living in a field dependent on oil prices that could—and did—fluctuate dramatically. As a result and quite understandably, they had significant issues, all of which went way over the head of a toddler preoccupied with playing in the snow and learning to crawl and to walk. The only thing I do remember about my early years living in Canada was that it was cold, though I doubt that was the only explanation for my parents' drinking.

Although my mom and dad had both been raised in nominally Christian homes, faith at that time was not real to either one of them. Neither one had a personal relationship with Jesus. My dad lived a fast life and was enjoying business success. When I was three he decided he no longer wanted to be married. And so he left my mother and me in December 1974.

I don't remember the fear and isolation that my mother must have felt. I don't recall the increased drinking that surely resulted. At the time, I was only a toddler, and so was insulated from the brunt of what was happening.

A few months later, back in Texas, a friend from the oil business invited my father to attend a Bible study being taught by a local life insurance agent. Although reluctant, my father decided to go, and he was struck by the peace he saw in the lives of those at the Bible study, despite the challenges they faced. One woman described how she lived with her son, who beat her to get money to buy drugs, and yet she was filled with what Scripture describes as "a peace that surpasses all understanding." My father was baffled.

The hosts gave him a copy of a booklet published by Campus Crusade and titled *The Four Spiritual Laws*. He read it carefully, and returned to the Bible study the next week.

He still had many questions, and they invited him to come to their home again the next night, to visit with Galen Wiley, their pastor from nearby Clay Road Baptist Church. For four hours, my dad argued with the pastor about the Bible, about religion in general, and about Christ. Finally, at about 11 p.m., my father asked, "What about the man up in the mountains of Tibet who has never heard of Jesus?"

Pastor Wiley wisely didn't take the bait. Instead, he replied, "I don't know about the man in Tibet. But you have heard about Jesus. What's your excuse?"

The question hit my father like a sledgehammer. And, shortly after 11 p.m. on Tuesday, April 15, 1975, he dropped to his knees and surrendered his life to Jesus. That day changed his life, and mine as well.

The following Sunday, he made a public profession of faith at Clay Road, a small church in the suburbs of Houston. And the next week, he went to the airport, bought a ticket, and flew back to Canada, returning to my mother and me. He asked my mom to forgive him, and for them to start over. Five years later, in 1979, I too asked Jesus to be my savior at Clay Road Baptist Church.

When people ask if faith is real, I don't have to speculate; I've seen the fruits of a walk with Christ in my own life and in my family. Were it not for my father's becoming a Christian, I would have been raised by a single mother. I would have lived without my dad in the home, facing the hardships that are unfortunately far too common in our society, as single parents so often struggle mightily to provide for their children.

Instead, my parents reunited. Shortly thereafter they sold their business in Calgary* and moved us down to Houston, where my mother also became a born-again Christian. Both of them quit drinking, and their lives were transformed.

---

*In 1975, my parents sold their small company, which had been renamed Veritas. Today Veritas has grown into a multibillion-dollar company. Alas, they retained no proprietary interest, and so they did not benefit from its subsequent growth.

––––––

Houston was a booming oil town when we arrived, and it was where I spent the remainder of my youth. For grade school, I attended West Briar, a small private school that had been founded by a number of Jewish doctors. It was a terrific school, and I was fortunate to have a fantastic teacher, Miss Jennings, for both third and fifth grades. She taught more grammar in elementary school than you'll find in most high school English classes (it seemed we would diagram sentences endlessly). Roughly half the school was Jewish, which led me to believe until I was ten that half the world was Jewish. Every year, we'd play with dreidels, enjoy latkes, and celebrate Hanukkah and Christmas side by side, and think nothing of it.

I went to the Awty International School for junior high. Awty, located in Houston, was half French; many of the students were the children of French diplomats or businesspeople. It was a strong academic institution, but I clashed with the teachers, strict disciplinarians whose European pedagogy consisted of them lecturing on and on and students silently taking notes. Instead, I preferred more engagement.

In both elementary and junior high, I was a geeky kid. As the child of two mathematicians, schoolwork always came easily for me, and I was consistently at the top of my class, except in handwriting (to this day, I still write in block letters rather than cursive).

My parents are both driven, and I share their competitiveness. Indeed, my dad and I both love games, from dominoes to cards, and when we played Monopoly my mom would flee the room laughing, it was so cutthroat.

Being hypercompetitive did not help, however, in sports. Neither of my parents is athletic, and I inherited their lack of talent in that arena. And since I wasn't very good, I refused to play sports as a child. That, of course, made me even worse. That mix—excelling in the classroom, being too competitive and cocky about academics, and being lousy at sports—was, needless to say, not a recipe for popularity.

Midway through junior high school, I decided that I'd had enough of being the unpopular nerd. I remember sitting up one night asking a friend why I wasn't one of the popular kids. I ended up staying up most of that night thinking about it. "Okay, well, what is it that the popular kids do? I will consciously emulate that."

First off, I decided that my existing policy of refusing to play sports simply because I wasn't good at them was not a wise plan if I wanted to be accepted by kids at school. I then decided to join the soccer team, the football team, and the basketball team. I was terrible at all three, but I kept at it. Around that time I got my braces off. I went to a dermatologist, and my acne cleared some. I got contacts instead of glasses. I also shot up about six inches.

I started trying to behave differently. I tried to be less cocky. When I received a test exam back, even though I'd probably done well, I would simply put it away. I wouldn't look at it. It wasn't rocket science, but it was interesting to see what these sorts of small conscious changes could produce.

Another thing that changed was my name. In Spanish, the diminutive is formed by adding -ito; thus, the diminutive of my full name, Rafael, was Rafaelito, which in turn was shortened to Felito. Until I was thirteen, I was "Felito Cruz." The problem with that name was that it seemed to rhyme with every major corn chip on the market. Fritos, Cheetos, Doritos, and Tostitos—a fact that other young children were quite happy to point out.

I was tired of being teased. One day I had a conversation with my mother about it and she said, "You know, you could change your name.

"There are a number of other possibilities," she said. And she proceeded to list them: Rafael. Raph. Ralph. Edward. Ed. Eddie. "Or you could go by Ted." I found that a shocking concept. It had never occurred to me that I had any input on my name.

"Ted" immediately felt like me. But my father was furious with the decision. He viewed it as a rejection of him and his heritage, which was not my intention.

"What do you mean Ted is a nickname for Edward?" he snapped at my mother. "Who's ever heard of that?"

My mother's response was unfortunate. "Well, there's Ted Kennedy," she said.

My father was apoplectic. He had no love for liberals. In fact, he believed the American far left was trying to turn this country in a dangerously socialist direction, much like the reviled Castro regime. One of the biggest fights he had had with my mother was in 1976, when she had voted for Jimmy Carter. (She quickly came to regret that decision when his haplessness became manifest.) To equate me with Teddy Kennedy was too much. For about two years, he refused to utter my new name.

My parents also decided to transfer me to Faith West Academy in ninth grade. Faith West was a brand-new school, located in a renovated "Handy Dan" hardware store. It was not at all academically rigorous, but my parents very much valued a strong Christian education.

At Faith West, I achieved something that had long eluded me: I was relatively popular. Again I played football, soccer, and basketball; I was still lousy at them, but had gotten marginally better. I also joined the yearbook, the newspaper, and the speech team, and ended up being twice elected class president.

Going from an unpopular kid in junior high to being elected class president in high school was, as one would imagine, a fairly startling transformation. It was fun. And, interestingly, it taught me a vital lesson: that popularity wasn't all that consequential. Happiness doesn't come from popularity, but rather from doing something that matters, making a difference, and fulfilling God's plan for your life.

Probably the biggest academic influence on my life in high school, and well beyond, for that matter, came from my involvement in what was then called the Free Enterprise Education Center (now known as the Free Enterprise Institute). The center was founded by Rolland Storey, a retired businessman and motivational speaker who had worked for Houston Natural Gas.

Mr. Storey was a short, impish man, balding with wisps of white hair. He was charming and irascible, reminiscent of commentator Andy Rooney on *60 Minutes*. He also was one of the most gifted natural speakers I've ever seen.

He had started a speech contest for high school kids based on "the ten pillars of economic wisdom," the first of which was "Nothing in our material world can come from nowhere or go nowhere, nor can it be free: Everything in our economic life has a source, a destination, and a cost that must be paid." That led directly to the second pillar: "Government is never a source of goods. Everything produced is produced by the people, and everything that government gives to the people, it must first take from the people." If only politicians in Washington understood that basic truth. Students were required to prepare a twenty-minute speech on all ten pillars, after reading a curriculum of economic fundamentals including the works of Milton Friedman, Friedrich Hayek, Adam Smith, Frédéric Bastiat, and Ludwig von Mises.

This was my first systematic exposure to free-market tenets, and it became the intellectual foundation for many of the ideas I instinctively believed: first and foremost that the greatest engine of prosperity and opportunity the world has ever seen has been the American free enterprise system. Such a system allowed someone like my father, who had nothing but a hundred dollars sewn into his underwear, to build a small business and achieve the American dream.

His opportunity to do that, to earn a living and prosper based on his own individual talents and hard work, was transformational in our family; that journey was commonplace in America, but extraordinary in the annals of history. Until the time of the American experiment, much of human existence had been, as Hobbes famously observed, "nasty, brutish and short." Now, with American free enterprise, the possibilities were endless—not guaranteed, but also not limited.

I relished the material and was fortunate to be one of the city winners of the speech contest all four years of high school. For that, we earned scholarship money, but even more important, Mr. Storey booked the contest winners to speak on free-market principles at Rotary, Kiwanis, and Exchange clubs and Chambers of Commerce all over Texas.

Standing in front of several hundred businesspeople to talk about free-market economics at the age of thirteen or fourteen was one of the most terrifying experiences of my life. In fact, it makes standing in the well of the U.S. Senate seem quite mild by comparison. Only adding to the tension when I started was the fact that I wasn't very good. The speech I had written had decent content, but my speaking style was stilted. To Mr. Storey, the solution was simple. The key to public speaking, he told me, was to pretend it was the most normal thing in the world. The very best speakers were those who can be in front of a thousand people and yet it feels as if they're sitting in a café chatting with you over a cup of coffee.

But acting "naturally" can be an incredibly unnatural thing, especially when you are standing in front of hundreds if not thousands of strangers. There is a reason that public speaking typically tops the list of people's greatest fears, ranking far higher than even death. But I practiced and practiced—with my father relentlessly critiquing my performance—and over time steadily became better.

In my sophomore year I also became involved in a spin-off of the program, called the Constitutional Corroborators. That consisted of five high school students who spent hundreds of hours studying the U.S. Constitution. We read the Federalist Papers, the Anti-Federalist Papers, and the Debates on Ratification. We then memorized the provisions of the Constitution in shortened mnemonic form.

For example, we memorized "TCC NCC PCC PAWN MaMa WReN." Those letters stood for eighteen enumerated powers of Congress in Article I, section 8 of the Constitution: "taxes, credit, commerce, naturalization, coinage, counterfeiting, post office, copyright, courts, piracy, Army, war, Navy, militia, money for militia, Washington, D.C., rules, and necessary and proper." If it's not in that list, Congress has no constitutional authority over it. (As I've often joked, you'll notice there's no "O" for "Obamacare.")

Twice every week, my mother would drive me forty-five minutes across town, to the public library in Pasadena, Texas (an industrial suburb of Houston); the five of us students would spend two hours studying with Mr. Storey, while my mom would quietly read a book. And then she'd drive me home.

After many months, we began touring the state of Texas, once again speaking at Rotary clubs and Kiwanis clubs. While our audience sat there having lunch, the five of us would set up easels in the front of the room. On the easels, each of us would write from memory the entire Constitution in a shortened mnemonic form. We'd write a definition of socialism ("government ownership or control of the means of production or distribution in an economy"), under the principle that if you don't know what it is, you can't recognize when you have it. And we would write a quote from Thomas Jefferson: "If a nation expects to be ignorant and free, it expects what never was, and never will be."

Then we delivered a thirty-minute presentation on the Constitution. We closed our presentation with a patriotic poem, "I Am an American," set to rousing music.

For each speech, the Center would credit us fifty or a hundred dollars in scholarship money. By the time I'd finished high school, I'd given somewhere around eighty speeches, so I'd earned a fair amount of scholarship money for college that way.

The more I studied free-market economics and the Constitution, the more obvious it became to me that there is a systemic imbalance in our political discourse. On the left, for the advocates of government power and statism, the best and the brightest are consistently attracted to the organs for the transmission of ideas, namely politics, journalism, entertainment, or academia. In that respect, Barack Obama is the epitome of the perfect leftist.

On the right, however, for those who believe in free markets and individual liberty, the best and the brightest tend to go make money. They go to the business world, rather than fight in the political arena.

Much of where we are today can be traced to that imbalance.

For me, I was likewise headed in the business direction. As the son of two computer programmers, I was sure where I was going. But then, when I was in tenth grade, someone asked me, "What do you want to do when you grow up?"

I repeated by rote what I had said many times before. "I want to go to MIT, and study electrical engineering. Then, get a Ph.D.

in computer science, and start a business designing artificial intelligence."

My father was standing there listening. He looked at me and said, "No you don't." Just very matter-of-fact.

I said, "What do you mean, 'No I don't'?"

"You haven't turned your computer on in six months," he replied. "All you talk about now is law and politics."

He was right. From then on, if you had asked me what I wanted to do in life, I would have told you, "do everything I can to defend free-market economics and the Constitution."

Admittedly, it was a bit of a strange answer (I was kind of a weird kid), but the Free Enterprise Institute—combined with my father's life experiences fleeing oppression and seeking freedom—helped me realize where my passion lay.

Even though I hoped to eventually run for office, I was convinced I would start off in business. I got my first job when I was ten years old, working as a computer operator for my parents' company. It was 1981, and the computer age was in its infancy. I would operate Raytheon 704 and 706 minicomputers, although there was nothing "mini" about them.

They were the size of a wall. They used tape readers and card punches, and they had less computing power than an iPhone today. For my labors, my father paid me a dollar an hour. (He didn't quite understand the child labor laws.) And he would sometimes assign me to work double shifts on Thanksgiving and Christmas Day, because no one else would work then. It was a small business and he didn't want to shut down for the holidays. (He was always nice about it; he'd bring me a plate of turkey and dressing, and then say, "Get back to work!")

My parents also insisted that I save 50 percent of what I made (a terrific life discipline to instill in children). I still remember the first thing I bought with my working wages: a Yorx stereo, which had a radio, record player, cassette, and eight-track player. In 1981, it cost $137, which represented 274 hours of work!

Three years later, when I was thirteen, I came into my dad's office to negotiate my first raise.

"Dad, you're paying me one dollar an hour," I said. "The other computer operators who work here, you're paying five dollars an hour. They're all college students. You're exploiting me because I'm your son."

My dad laughed. "I was waiting for you to figure that out," he said. He promptly gave me a raise to five dollars an hour, which I supplemented by stocking the office's Coke machine with fifteen-cent cans bought in bulk from the grocery store and later by running a small lawn-mowing business.

In the summer of my sophomore year, I decided to start a business under the name Cruz Enterprises, which I'd borrowed from the multibillion-dollar "Stark Enterprises" in *Iron Man* comics. As a teenager, I figured I could do a lot worse than trying to model my business after Tony Stark's.

I had spent the previous summer going door-to-door, setting up leads for salespeople to come by later and try to sell water softeners. The company would give folks $20 in free groceries for taking the meeting, and they'd pay me $15 for every meeting I set up.

My new business model was simple: I was going to hire other kids to join me knocking on doors. I'd pay them $5 an hour, plus a small production bonus. And then I'd pocket the difference between the business costs and the commission we'd earn. I advertised, running an ad in the "Green Sheet" classifieds, and hired a couple of other kids.

My first two employees were brother and sister. And, sadly, my business model was not destined for success. Neither of them worked very hard. They would lackadaisically knock on doors, and they set up only a few leads. I kept paying them, but they were generating very little. By the end of the summer, I had broken even; the leads I had generated myself just paid for the hourly wages I had committed to my employees.

At one point their mother began yelling at me, "You're working them too hard! It's unreal how hard you're making them work!"

"But they're doing the same thing I'm doing," I replied. "They're knocking on doors on one side of the street. I'm on the other."

She was furious that I would expect actual work from her children. It was a good lesson that not everyone shared the work ethic my parents had endeavored to teach me.

———

I guess it's fair to say that I was always a driven individual—but that's not to say I always went in the right direction. Sometimes I drove myself and others right into trouble. For example, in high school I was suspended from school for several days for going to a party, drinking, and smoking pot (something I experimented with as a teenager before wising up).

My father was convinced that I had just gotten in with bad kids who were leading me astray. My mother, on the other hand, held me firmly responsible for my own actions. "Apparently you don't know your son very well," she said. "I'm sure he was leading the effort for whatever happened."

My best friend in high school lived two houses down from me. On the surface, Dwight Odelius and I could not have been any more different. He was an artiste, a gifted pianist and fiction writer who attended one of Houston's magnet schools, the High School for the Performing and Visual Arts. A year and a half older than me, Dwight had long, floppy hair, wore lots of black, and could ride a unicycle.

He was razor-smart, and something of a hippie. Although most of his friends were politically liberal, Dwight was not. He had been adopted at birth, by wonderful parents who gave Dwight a nurturing home, and so being pro-life to him was always very personal. He was immensely grateful that his very young birth mother had chosen to give him up for adoption, to allow him to live, rather than making another choice.

We both lived on a small lake on the outskirts of Houston, and Dwight and I would take our rowboat onto the lake at night to catch bullfrogs. Doing so is simple: You shine a light along the shore, and their eyes glisten in the light. As long as you keep the light locked on the eyes, the frog won't move. And then you use a frog gigger (a long pole with a clamp on the end) to catch them.

We'd also mess with the occasional alligator, and once caught a live cottonmouth snake, which didn't make Dwight's dad very happy.

Dwight and I got into a fair amount of trouble. We set up secret intercoms connecting our houses, and we'd both sneak out late at night, push his car silently down his driveway, and head to parties, doing our very best to pick up girls. Sometimes we were successful, other times not so much. We'd also pull teenage pranks, including at Christmastime taking the lights from several neighborhood houses and then surreptitiously decorating another house with them.

Sometimes the consequences of my teenage hijinks were more serious. Sophomore year in high school, I went to spend the night at my friend David's place. David's family went to our church, and his parents really struggled to pay the bills. At night, the two of us used to sit in the dark living room of his apartment with a pellet gun and shoot the rats that would come out in the kitchen.

That evening, at about midnight, David and I snuck out the window to head to an arcade down the street. We hung out and played foosball until it closed at 2 a.m. At that point, we were standing in the parking lot debating what to do next when a car of four boys drove up. David and I were both fifteen, and all four of the boys were older, probably seventeen or eighteen. They were drunk, and I guess the two of us looked like easy prey.

The four of them jumped out of the car, walked up to us, and proceeded to beat the heck out of us. One of them bent my arm behind my back, slammed my head onto the back of a car, and began punching me repeatedly in the face. My shoulder was nearly dislocated, and one of my contact lenses ended up being knocked out of my eye, ending up somewhere on that car.

Fortunately, they were drunk enough that David and I were able to twist ourselves away. He ran in one direction and I ran in the other. I went to David's house, and when he didn't return, I ended up waking up David's parents and telling them what happened.

David had run to my sister Miriam's house a few blocks away. Miriam was my half sister, the eldest of my father's two daughters from his first marriage. Miriam's husband, Joe, was a difficult character who had spent much of his life in and out of jail. When David woke them up to tell them I was missing (he thought they had taken me and thrown me in the car), Joe got a shotgun and a baseball bat and took off in his car looking for me.

Things could have gotten ugly. Eventually David called home and talked to his parents, and everything was sorted out.

I spent the rest of that night at Miriam's. The next day I walked over to my parents' office, which was next door. When my father looked up he saw my black-and-blue face. "What happened?" he asked.

I recounted the tale. He thought about it a moment and said,

"Well, I guess you learned your lesson." And he looked down and went back to work.

My mom had a different reaction. She broke into tears, scared for what might have happened.

Between the two, my dad's reaction was much easier to handle.

Midway through junior year, I transferred to Second Baptist High School, which had far more rigorous academic standards than Faith West. It was a small Christian school; my graduating class was just forty-three students.

In my senior year at Second Baptist, I again tested the authorities. Our school's archrival was Northwest Academy. On the day of our homecoming, the basketball players at Northwest came to our school and stole our school flag.

Outraged, I assembled a group of three other students to avenge this wrong. We drove to Northwest Academy in my car, a 1978 green Ford Fairmont given to me by my grandfather. We called it the Green Bomb. And it was a wonderful car to give a teenage boy because it was basically a tank, a very large hunk of metal that got about ten miles to the gallon.

That night my buddies and I picked up thirty-six rolls of toilet paper, several cans of shaving cream, toothpaste, and baby shampoo and then we wrapped the entire building of Northwest Academy in toilet paper. With the shaving cream, we drew funny faces on each of the windows of the school bus. We even brought a thank-you card, which we signed in red lipstick with the words, "It's not nice to steal," and managed to make our way into the gym to leave it in the center of the basketball court. All in all, it was rather mild stuff as pranks went. Or so I thought.

When we were done, a number of janitors who were cleaning the place spotted us and began running after us. We jumped into the Green Bomb, and I took off driving like a bat out of hell, tires screeching, racing through the neighborhood as fast as I could. The janitors got in a car, too, and started chasing us and yelling out of their windows.

As it so happened, the car stereo was playing my friend Joel's tape of "Ride of the Valkyries," Wagner's fast-paced theme. To have the anthem of *Apocalypse Now* playing during a car chase made things pretty surreal.

I then decided to shut off the lights to make it harder to follow us, which was really dumb because it was also now harder to see. At one point my car jumped a curb, landed on someone's front lawn. and hit a fire hydrant. Now in a panic, I pulled out across these poor people's front lawn, trenching it with my tires. I still couldn't lose the car behind us.

Finally I came up with what I believed to be another clever plan. "Okay, we're going to stop," I told my nervous passengers. "When we stop, they'll stop, too. Then when they get out to come after us, we'll slam on the gas and take off again."

Everyone seemed to think this was a good plan. So we stopped. As expected, the other car pulled up behind us. Then we waited. And waited some more. No one exited the other vehicle.

"What are they doing?" one of my friends asked.

Peering at them through my back window, I realized what a mistake I'd made. The janitors were writing down my license plate number. And then they left.

The next day I returned to my school. We had a big pep rally planned that day, and while all of the students were gathered, they made an announcement. I was to report to the principal's office.

Principal Doss looked at me from across his desk. "Ted, do you drive a '78 Ford Fairmont with license plate such-and-such?" He was using an old attorney's trick—never ask a question to which you don't already know the answer.

Everything he was doing was designed to intimidate me. But even then I wasn't someone easily intimidated by those in authority. I also tried very hard to be honest.

"Yes, sir, I do," I replied.

"Were you by any chance at Northwest Academy last night?" Another question whose answer he already knew.

"Yes, sir, I was."

"Okay," he said, leaning forward. "Tell me who was with you."

There was no way I was ratting out my friends.

"I'm sorry, sir, I'm not going to do that," I said. "I'll admit to what I have done, and I'll face the consequences, but I'm not going to give you the names of anyone else who was involved."

He seemed to expect that response from me. "All right then," he said. He then reached into the drawer of his desk and pulled out a letter that he had already typed and signed. The letter was addressed to Princeton University recommending that my admission to the college be rescinded.

This was an eye-opener for me. Just a few months before, I had been thrilled to be accepted to Princeton. For a boy whose father had washed dishes for fifty cents an hour, going to Princeton was a dream come true. No one in my family had ever imagined going to an Ivy League college. When I received my acceptance letter, I couldn't believe I was going to get to go to the school where F. Scott Fitzgerald studied before writing *The Great Gatsby* and where George Kennan studied before writing the "Long Telegram," which set our country on a course for winning the Cold War.

But now I was faced with an altogether different letter—one with the potential to keep me away from the school I had worked so hard to attend.

"Do you want me to send this letter?"

I answered truthfully, "No, sir, I do not."

"Well then, tell me who was with you."

I wasn't sure if he was serious, but I wasn't going to budge in any event. "Sir, I've already told you, I think it would be in poor character and contrary to my integrity to give you those names, and I'm not going to do so."

He said, "Fine, then we're calling your father."

"Okay. Here's his number." I wrote it on a piece of paper.

He reached over to his telephone and dialed my dad, and described his version of what had transpired. Then he handed the phone to me.

"What's going on?" my father asked. "Look, you've got three months left; just graduate from high school and go to college."

I said, "He's asking me to give the names of others who were involved in this. I won't do it. It just isn't right."

My dad thought about that for a moment. I hoped he'd say the right thing, which, of course, he did. He said, "Well, stick to your guns, son. I'm proud of you."

I told the principal I wasn't going to change my mind, and that he could send that letter if he wanted to (although it occurred to me that Second Baptist would probably not really want to force the rescission of the first Ivy League acceptance in the school's short history). He put the letter back in his desk and called in an assistant. "Go get Joel, Luke, and Patrick," the other three boys who were with me.

Clearly, the principal already had his suspects. As it turned out, and I subsequently learned, one of the other three had confessed earlier to the basketball coach.

The whole effort in the principal's office then was simply a power play to get me to buckle. He knew the answer to everything he'd asked me, but he was determined to have me rat out my fellow students to teach me a lesson.

When my three conspirators came into the office, the principal asked all of them if they were there with me. Each replied, "Yes."

The principal then began to chew each of us out, for our immaturity and putting the school in a bad light. Our basketball coach was also in the room, watching.

"Coach, is there something you want to say?" the principal asked the coach, once he took a breath.

"Sure," he replied.

"You shouldn't have done it," he said, looking intensely at the four of us. "But I'm sure glad you did," he said laughing loudly. The principal just glared.

The saga ended with each of us going to the principal at Northwest Academy and offering effusive, over-the-top apologies. They were not entirely sincere.

————

My brushes with the authorities did not make things easier on my parents during what was a difficult time for them. My dad had been dealing with the slow unraveling of the company into which he and my mother had put their heart and soul since returning to Houston.

My parents' company was supposed to bring together their unique skills to create a niche for themselves in the oil business. To search for pockets of oil, companies used dynamite to send out sound waves underground. Those waves bounced off various densities of rock below the surface and sent signals back to the surface, creating a chart of numbers that could be deciphered into a virtual map of vast underground terrain—showing where there were rocks and where there were pockets that might contain oil or natural gas. Often the data coming back were not easy to decipher, so my parents would run it through their customized computer programs and print out what looked like a slice of the earth. (Back then, it was all two-dimensional; today, 3-D seismic mapping is the industry norm.) A self-trained geophysicist, my dad analyzed these maps and assessed which pockets were potentially the most fruitful for exploration.

In the beginning they were quite successful. At its peak the company had had about twenty-five employees. And for a number of years, when oil prices were high and oil companies were flush with money and looking to expand, my parents' company did well.

When business went well, my parents put their energy into church and Christian charities. My dad served on the board of the Star of Hope, a wonderful nonprofit that ministers to and cares for the homeless in Houston and helps them get back on their feet. He also served on the Texas board of the Religious Roundtable, a national organization that helped mobilize millions of people of faith to turn out and elect Ronald Reagan president in 1980.

Just a few years ago, I learned something new about that time. My dad was in town and I brought him to our church, First Baptist Church of Houston. After church, I was introducing him to several friends, and one of them, a deacon, recounted a story:

"Ted, I actually met your dad thirty years ago, although he probably doesn't remember it. At the time, I was a young associate at a law firm, and he hired the firm to give him tax advice because he wanted to give away over fifty percent of what he was earning. It made a real impression on me."

My dad had never told me that.

For a time, business was good. But then, in the late 1980s, circumstances beyond my parents' control conspired against them. Ironically, President Reagan had an inadvertent hand in their company's demise since one of the ways he won the Cold War was by strengthening the dollar, which caused the price of oil to plummet. As Reagan intended, this had a crippling impact on the Soviet

Union, which depended on the sale of oil abroad to sustain its economy. But the policy also had a severe domestic impact, particularly in Texas, and most of all in Houston, which at that time was very much a one-industry town.

When oil collapsed from $30 a barrel to less than $10, the city was devastated. Whole skyscrapers were all but vacant. Clients that used to regularly hire my parents' company for tens of thousands or hundreds of thousands of dollars a month, overnight went down to zero.

The oil business has always been cyclical, and so my parents tried to ride out the downturn. They tapped their personal savings and funneled it all into the company, but to no avail.

On one somber Monday morning, my dad was forced to lay off 19 of his 25 employees. These were people who had spouses and kids and mortgages. When Dad came home that night, he looked like he had been beaten with a two-by-four. I've never seen him so unhappy. What made matters even more poignant was that some of his employees were so loyal to him they refused to leave. They would say, "Raf, we're not going anywhere. We're staying here with you. We can make this work." He had to argue with them. "Listen, you've got a wife. You've got kids. I don't have the money to pay you. You have to go somewhere else to feed your children."

Ultimately the company went broke and my parents went bankrupt. It was a full-bore disaster. Whatever other money my parents had had been invested in real estate, but when oil prices plummeted, so did Texas real estate prices. We lost our home and had to rent a smaller and cheaper place in the suburbs. To keep us afloat, my dad worked job to job—as a salesman, a financial planner, whoever was hiring. But my parents were in their fifties, which made it even harder for them to find jobs.

In my mother's case, she was too experienced for most jobs

she applied for. She ended up stripping almost everything off her résumé—even her Rice degree—so she could get hired in an entry-level job just to pay the bills. She had worked so hard in her life to achieve in a male-dominated business, and now she had to start all over again. Everything she had struggled to do for so many years was now a hindrance.

As their business crumbled, it put enormous stress on their marriage, so much so that a few years later, their marriage fell apart. That was particularly hard for me to accept. Many young children of divorce sadly blame themselves. When my parents split, I was an adult in law school, so fortunately I understood it was not because of me.

But I tried hard to keep them together. I argued, I cajoled, I pleaded. I printed out and gave my Dad pages and pages of Scriptures on the sanctity of marriage. But, alas, there was nothing I could do. I had to accept that they were adults—like all of us, troubled, flawed, and human—and they were entitled to make mistakes. Even if it cost them their marriage.

Another consequence of their bankruptcy was that they could no longer help me pay for Princeton, which made me financially independent (involuntarily) at the age of seventeen. When it came to college, I was on my own. Fortunately, I had the money I had earned making speeches and had received a number of other scholarships that helped defray some of the costs. I took out need-based financial aid and worked campus jobs to cover the rest.

★

CHAPTER 3

Enter the Ivies

She was a junior senator sitting across from twelve officers—the most senior uniformed men and women of the U.S. military, including the chairman of the Joint Chiefs of Staff and the chiefs of the Air Force, Army, Navy, Marines, and Coast Guard. And that didn't bother Kirsten Gillibrand in the least.

The June 4, 2013, hearing of the Senate Armed Services Committee was not convened to discuss matters of war or peace, but an epidemic from within, one damaging the culture of discipline within the military ranks. It included an unprecedented showing of military brass in the halls of Congress. The officers' combined thirty-nine stars gleamed in the hearing room's unforgiving lights and flickered in the constant flashes from cameras.

Though the vast majority of the men and women in uniform are capable and honorable, the U.S. military had handled sexual assault and harassment cases the wrong way for many years. In 2012, there were more than 26,000 instances of unwanted sexual contact in the U.S. military. Fewer than 3,400 of them were formally reported through the chain of command. Unfortunately, the brass didn't have a viable plan to address this growing crisis, one that imperiled the military's performance, other than something approximating "Trust us. We'll get it right this time."

That wasn't enough for the determined Democrat who had assumed the New York U.S. Senate seat left vacant after Hillary Clinton resigned to become secretary of state. Gillibrand had built a reputation for bucking the standard way of doing things—she first won a seat in the House of Representatives by running as a Democrat who supported Second Amendment rights and opposed amnesty for illegal immigrants. Once elected, she became the first member of Congress to publish her entire official schedule every day, so that her constituents knew exactly with whom she met, including lobbyists and lawyers. She also made a point of publishing the earmark requests she had requested and received, as well as her own personal financial statement. As a senator, she authored parts of the STOCK Act, which imposed limits on the scandalous but legal practice of insider trading by members of Congress. That didn't win her any new friends among her colleagues. Neither did her determination to confront the senior leaders of the U.S. military, as she was doing today.

"You have lost the trust of the men and women who rely on you—that [you] will actually bring justice in these cases," she told the assembled officials. "They are afraid to report. They think their careers will be over. They fear retaliation. They fear being blamed."

An accomplished lawyer, she had been working for months on legislation to reform the military justice system. In keeping with tradition of the U.S. military, the decision to refer cases for prosecution of a crime—whether refusing an order or murder—lies solely with the commanding officer. That is one reason that tens of thousands of cases of unwanted sexual contact go unreported in the U.S. military every year. Victims are afraid to report the crime to their superiors, who are the ultimate arbiters of whether the offenders will be prosecuted or not. In the most flagrant cases, those arbiters are the offenders themselves.

In these efforts, Gillibrand stood against not only the entire senior rank of the Department of Defense, but also the presidential administration of her own party, as well as the powerful chairman of the Senate Armed Services Committee, Senator Carl Levin, and dozens of other Democrats unwilling to challenge the status quo or pick what might be an unpopular fight against the military.

I knew Senator Gillibrand was right in her approach, which is why I signed on to the bill as a cosponsor. The truth is that too many victims fear that their commanders cannot be objective about the men and women in their command. Nor can they know the ins and outs of sexual assault, as an experienced military prosecutor does. Israel, the United Kingdom, Australia, and Germany have made reforms similar to the ones Gillibrand put forward. The result was a marked improvement in how cases of sexual assault are reported and prosecuted.

But the issue was profoundly personal for me as well. I knew what my aunt had experienced when she was a prisoner in Fidel Castro's Cuba. And I had been an advocate for enhanced sexual assault prevention efforts while in student government at Princeton.

It took almost a full year after that hearing for Senate Democrats to allow Gillibrand's bill to come to a vote. Even then, as I fought alongside Senator Gillibrand on March 6, 2014, it fell five votes short of the number needed to overcome a Senate filibuster. In public remarks after the vote, Gillibrand publicly faulted her own party's president, Barack Obama, for failing to support the measure.

That June, a military court-martial finally disposed of one of the most notorious sexual assault cases in memory. General Jeffrey Sinclair, a twenty-seven-year Army veteran, had been accused of sodomizing a female soldier and threatening to kill her if she told anyone about their affair.[1] The charges carried a maximum jail sentence of twenty years; by most accounts, General Sinclair got off easy. He was spared any jail time and fined twenty thousand dollars. Gillibrand was outraged. "This case has illustrated a military justice system in dire need of independence from the chain of command," she said. I knew she'd press on to fight the good fight again. And in time, we'll change the system.

———

I was thrilled to go to Princeton. The first time I saw the campus, I fell in love with it. It was idyllic, picturesque, and looked like what an elite college campus is supposed to be. The Gothic architecture, majestic and covered in ivy, took my breath away. Countless black squirrels playfully chased each other in circles on the central lawn. It was also hard not to be impressed by the fact that, in revolutionary times, the Continental Congress had met at Nassau Hall, the central building at Princeton.

In the summer after my junior year of high school, my parents and I had toured about a dozen schools—Georgetown, the Massachusetts Institute of Technology, Harvard, Dartmouth, and others. Each time I would ditch my parents as soon as possible and go and speak to actual students. I asked things like, How do you like this school? What's it like? More than any other place we visited, it struck me that Princeton students loved to be at Princeton.*

---

*My second choice would have been Dartmouth, which was an outstanding institution with an incredible campus in the rather isolated town of Hanover, New Hampshire. The students were very welcoming. On my visit there, I spent the night with some Dartmouth students, who brought me to a frat party. What followed was an evening of dice games and significant quantities of beer. Much to my parents' displeasure, when they picked me up the next day for our trip to Brown University, I was not in an ideal frame of mind. In fact, at a meeting the next day with an admissions counselor at Brown, I had to ask her

Princeton was a completely new world for me, and it wasn't easy to adjust. Most of my fellow students had come from elite schools, whereas back home I knew virtually nobody who had gone to an Ivy League school. As the son of a Cuban immigrant with a little bit of a Texas drawl and cowboy boots, I felt distinctly different, and I found out in short order that my roommate agreed with me.

Much to my dismay, my randomly assigned freshman year roommate—a liberal student from New Jersey—took an immediate dislike to me. I had never had a roommate before; my half sisters were much older than I, and so I was mostly raised an only child. I had hoped my roommate and I would be lifelong friends, but alas, he spent much of his time treating me with contempt. I have no doubt that, as an immature seventeen-year-old, I contributed to his antipathy (indeed, my annoying habit of repeatedly hitting the snooze button led him to surreptitiously superglue it in place), but it was nonetheless disappointing.

But I was rescued from an entire year of petty torment by the most unlikely of saviors.

About a dozen of us were gathered in our resident advisor's dorm room when a sixteen-year-old Jamaican student entered the room a few minutes late. David Panton was six foot three, over two hundred pounds. Handsome and graceful, he immediately went up to each student and in his distinctive accent warmly introduced himself. He told us he was born and raised in Mandeville, a small town "in the bush," as he put it. He radiated charisma and integrity. He almost instantly became my best friend.

We found in each other strengths that complemented our own individual weaknesses. He found in me an intensity and drive to

---

to please lower her voice because, I told her, I was really hungover. That probably did not leave the best of impressions.

succeed, which in turn helped him realize his own potential. And I found that he had a natural charm and ease with people, an EQ, if you will, that can only be called God-given and with which I had not been blessed.

I learned this the hard way when I was unexpectedly invited by members of the Princeton basketball team, who lived upstairs in our dorm, to participate in a regular game of poker. I had enjoyed playing poker as a kid. I had learned to play from my maternal grandmother when she babysat me. And I thought I was pretty good.

Granny was a wonderful card player; she had played bridge almost every week for sixty years. She taught me to play poker using buttons and costume jewelry as chips. I always won. When I was about eight, I came over one evening and told her, "Granny, let's play for real money." I had five dollars, which I'd saved up from ten weeks of allowance (fifty cents a week) and which I displayed to her proudly with every expectation of doubling or even tripling it. Granny said, "Sure." For the first time, she didn't let me win. She cleaned me out of my five dollars, and when my mom picked me up that night, there was a twinkle in her mother's Irish eyes. I was in tears, my pockets empty.

Alas, I didn't learn the lesson I should have from Granny: that famous admonition about a fool and his money.

At Princeton, I was so taken with playing poker with these popular basketball players that it never occurred to me that they didn't actually want to be my friends. They were sophomores, cool varsity jocks, and I was a "fish"—an easy mark. And, unbeknownst to me (and one other fellow), they had agreed to secretly pool their winnings, a huge advantage in a poker game. By the time I realized what had happened, I owed them two thousand dollars.

My mom and dad were bankrupt. My campus job barely covered my existing expenses. I was embarrassed and didn't want to admit my stupidity to my parents. So I called my Tía Sonia, who was work-ing in a bank at the time. She quietly arranged for me to get a loan of two thousand dollars. When I gave the money to the guys—in cash, forty crisp fifty-dollar bills—I told them that I knew they had suck-ered me out of the money. But since I had made the bets, I intended to honor my word. That left me with an outstanding loan that I paid back, on time, over the next two years; I took a second campus job to make the payments. It also left me with another important lesson about the perils of trying too hard to be popular.

My first job was filming and editing videos for the media services office of the university. It paid $7.50 an hour, and was pretty fun to do. My second job was teaching the SAT and the LSAT for the Princeton Review. I've always enjoyed teaching, and this job paid $15.00 an hour, which really helped cover my expenses.

It took me a little time to find my footing academically. My major was in the Woodrow Wilson School of Public and Interna-tional Affairs, commonly known on campus as "Woody Woo." It was a multidisciplinary major, which you could tailor to fit your interests. I ended up taking primarily economics and political the-ory and philosophy courses, all of which I really enjoyed. But unlike in high school, I was not a top student. Princeton was harder than Second Baptist (to put it mildly), and the first couple of years, I just didn't put in the work that I should have academically.

Instead, I had dived into a variety of extracurricular activities, especially debate and student government. Junior year, David was elected student body president and I was elected to the University Council, which I went on to chair. Together we led the student government. We fought back overly politicized efforts—such as the misguided drive to expel ROTC from Princeton—and tried to focus on more mundane matters, such as improving campus life for our fellow students. We worked to improve the student center, to enhance food quality in the cafeterias, to improve campus safety and rape awareness and prevention, those were the areas where we focused our energy.

My extracurriculars, along with my two jobs, consumed most of my waking hours, so much so that on the way to one of my final exams, I had to ask another student where our classroom was located. When I got my grades I found that my lackluster efforts were not up to Princeton's standards. The grades weren't terrible, but instead of the A's I'd gotten in high school, they were mostly B's, which could pose a problem for getting into law school, which I was beginning to envision as my next academic step.

I knew I needed to focus more on my classes, but I found it very difficult to curtail my interest in the debate team. I hadn't formally debated before (our high school was too small to have a team), but it was fascinating. The debate team came to consume an enormous portion of my life at Princeton, and it helped me hone a set of skills that proved immensely useful in my subsequent career. Debates at Princeton were far more complex and rigorous than anything approaching "debate" in a political campaign or on the floor of the U.S. Senate. The latter two, alas, more closely resemble an exchange of talking points—two ships passing in the night—rather than an actual give-and-take of ideas and arguments.

The debate team at Princeton is part of the American Whig-Cliosophic Society, the largest extracurricular group on campus. Originally two societies dating back to 1765, the American Whigs had been founded by James Madison when he was a student. The Clios were founded by William Paterson. In 1928, after the two societies had battled endlessly, they were forcibly merged by the university.

Madison, considered the father of the U.S. Constitution, is one of my personal heroes, and his genius in restraining government to allow individual liberty to flourish is what set the foundation for our prosperity. However, the party he founded, the American Whigs, had become the liberal party on campus, and the Clios were the conservatives. I ended up chairing the Clios, and David was my whip. I always found it ironic that I couldn't join the party Madison had founded even though I agreed with virtually every one of his views.

In one of our last debates leading the Clios, David and I argued the position that "Princeton should end affirmative action." Together we argued that, rather than discriminating based on race, Princeton should instead adopt economic affirmative action, targeting low-income prospective students. That policy would accomplish similar ends, but would be far more fair. It was striking to have that position advocated by a Hispanic man and a black man, and apparently we were persuasive; when the roughly one hundred or so students who attended the debate voted at the end, our side prevailed by a substantial margin.

At Princeton, debate was modeled after the British House of Parliament. In each round a participant would either represent Her Majesty's Government or the Loyal Opposition. The government would propose an affirmative case—such as "We should withdraw from NATO" or "We should eliminate the estate tax"—while the opposition, obviously, debated against that proposition. The parliamentary style of debate was meant to encourage argument based on values, principles, and reasoning. By our sophomore year, David and I had become regular debate partners, and between practice and competition, debating took up just about every weekend and much of the week.

Earlier I mentioned that one of the qualities David found appealing in me was that I was driven. Well, debate sometimes showed him the flip side of that coin. During a debate tournament, we typically drove to the tournament site (usually colleges up and down the eastern seaboard) on Friday morning. Then, that afternoon and evening we would have a couple of rounds, followed by a debate party. On Saturday we would have a couple more rounds, going to quarterfinals, semifinals, and the finals. Debates would usually wrap up late Saturday afternoons when we would drive back to Princeton, arriving on campus around midnight or so. In each round of the tournaments, judges prepared written ballots scoring individual speeches and critiquing the debate.

When we returned to Princeton on these late Saturday nights, I insisted that David and I go up to our room, sit down, and assess our performance so we could learn from our mistakes. We would spend hours reviewing the ballots. If we lost one round, for example, we would go on endlessly dissecting what we did wrong, and how we could have done better. Even if we had won a round, we'd ask how the other side could have beaten us. Or how we could have won by a bigger margin. We learned from these discussions, which would extend till three or four in the morning, which would often prompt David to protest, "Enough already! This is madness."

But the madness paid off. By senior year, David and I had debated in all four years of college. We had won dozens of tournaments. And we ended up winning "Team of the Year," awarded to the top debate team in the country. Individually, I was named "Speaker of the Year" and David was the named the second-place "Speaker of the Year."*

---

*The prior year, we had been named the second-place Team of the Year, beaten for the top spot by our friends and very talented Yale debaters Austin Goolsbee and David Gray. That year, Gray was named Speaker of the Year, and I was the second-place Speaker of the Year. Gray became an Anglican minister in Maryland. Goolsbee, who was a year ahead of us, would go on to become a well-regarded professor at the University of Chicago and the chairman of President Obama's Council of Economic Advisers.

Austin's economic views tend to the left, but he's one of the funniest people you could

Debate was the most rewarding experience of college. It was great fun, and it taught me a great deal.

However, by the time my junior and senior years came around, I became far more serious about my grades, realizing that I needed to bring up my GPA if I wanted to attend a top law school. I went to the financial aid office, and was able to get additional student loans. I wasn't thrilled about the debt burden, but the loans enabled me to stop working my campus jobs and focus instead on my studies.

The additional work yielded results. Junior year, my grades steadily rose from B's to A's, and even more so in senior year.

But there was one unfortunate exception. Princeton does a terrific job teaching students to write, both through incessant essays in class and also through requiring students to write three major academic works: a senior thesis and two junior papers ("JPs").

One of my JPs was overseen by a fairly well-known liberal professor on campus named Stanley Katz. My paper compared the benefits of private charitable institutions with government welfare, assessing which produced better results for the people being helped.

My paper concluded that, if you want to really help those in need, private charities do it better. The JP examined empirical data and also drew heavily from interviews I conducted, including with members of my church in Houston, which at the time ran a substantial food bank program and worked closely with the homeless.

Both reflected the reality that volunteers in the private sector who depend on donations to keep their efforts afloat have a vested interest in helping people down on their luck get back on their feet, so that the charity can then help other people in need. The best cure for poverty is not temporary food and shelter (although those are

---

ever know. Indeed, one time, when we were playing basketball against each other, he repeatedly taunted me (in a faux southern accent) to take shots farther and farther out. He was so hysterical that I couldn't resist, and they beat us at hoops, too.

certainly needed), but a job and the ability to provide for your family. As the adage goes, Give a man a fish, feed him for a day; teach a man to fish, feed him for a lifetime.

And private charities are far more likely to work not only to feed and clothe those in need, but also to help train them and get them interviews for jobs. Moreover, through the church, they can also help with their spiritual needs, which can be transformational in their lives.

Under government assistance, by contrast, there is far less of an incentive to help people become independent. Government programs don't tend to run out of money, regardless of whether they help people or not. In fact, the larger the homeless problem, the more money government programs receive.

When the paper was submitted, it received a B, which is a perfectly good grade. I'd received quite a few B's earlier at Princeton in classes that I rarely attended. But I'd worked hard on this paper, interviewed lots of people, and felt that it deserved a higher grade.

So I went to Katz's office, told him I was confused by my grade, and asked him what I had needed to do to get an A. He told me he was displeased that I had cited some of the empirical work on poverty done by Charles Murray, a conservative scholar whom I would later learn Katz despised. He looked at me dismissively and said, "I don't think you are academically capable of earning an A."

His answer took me aback, and at the time I wondered what was driving his condescension and hostility. Was it ideological prejudice, or something else?

Regardless, his assessment of my ability differed markedly from that of my other professors; indeed, based on the remainder of my departmental GPA, that single grade likely knocked me down from graduating summa cum laude to graduating cum laude.

Fortunately, that experience was the exception, not the rule. I had wonderful professors in college, most notably Robby George, one of the leading conservative thinkers in the nation. Learning from Professor George was one of the best things about Princeton. The *New York Times* has called him the "country's most influential conservative Christian thinker." As Glenn Beck has observed, George is "one of the biggest brains in America." From abortion to marriage to the natural rights of men and women, George is a sometimes lonely but always powerful voice within academia for the Judeo-Christian values on which this country was founded.

Professor George oversaw my other JP, and he then became my senior thesis advisor. We became friends, and I came to know his sense of humor, sometimes at my own expense.

When I received my JP back from Professor George, the top right corner of the first page was folded over. On it was written "C+." The blood literally drained from my face. With the super-seriousness of a nineteen-year-old, I was certain that my chances of getting into law school were finished. With white knuckles, I folded the corner over, and on the front was written "Just kidding! A."

Professor George found it pretty funny, as did I . . . years later.

My senior thesis focused on the Ninth and Tenth Amendments to the Constitution. The Ninth says that the "enumeration . . . of certain rights shall not be construed to deny or disparage other rights retained by the people." The Tenth says that the "powers not delegated to the United States by the Constitution . . . are reserved to the States respectively, or to the people."

My thesis argued that both amendments work together to restate the same principle, that we have a federal government of enumerated powers. Rights and powers are mirror images; a right is an immunity from government authority, and a power is the ability for government to act where rights end.

When the Constitution was adopted, it lacked a Bill of Rights.

The Framers engaged in a vigorous argument about whether that was a fatal flaw. My thesis retraced that argument. The Federalists argued that a Bill of Rights was unnecessary because the federal government couldn't violate those rights anyway. Why protect the freedom of speech, or religion, or the right to keep and bear arms, when restricting those rights didn't fall within the enumerated powers of the federal government? And if we added a Bill of Rights, it would suggest that otherwise the government could indeed restrict them.

The Anti-Federalists replied that politicians always want more power, so we should do everything we can to safeguard our rights. And our rights could be violated even within the enumerated powers; for example, the U.S. Post Office (explicitly authorized by the Constitution) could be run in a manner that censored our mail, violating our free speech.

Ultimately, the Anti-Federalists had the better of the argument, and we adopted a Bill of Rights. But the Ninth and Tenth Amendments, the last two of the Bill of Rights, were added to make absolutely clear that federal government's power remains limited. Listing rights that are protected doesn't imply that those are our only rights, or that the federal government has authority beyond that expressly given it.

The Framers had seen firsthand what happens when government power is unchecked; indeed, we had just fought a bloody revolution to free ourselves from a distant monarch. The Constitution was designed, as Jefferson put it, "as chains to bind the mischief of government."

In Federalist 51, James Madison explained, "If men were angels, no government would be necessary. . . . In framing a government which is to be administered by men over men, the great difficulty lies in this: you must first enable the government to control the governed; and in the next place oblige it to control itself." Drawing inspiration from that insight, my thesis was titled "Clipping the Wings of Angels."

As I completed my thesis, I also applied to law school. I thought seriously about going home, to the University of Texas School of Law. It's a terrific school (indeed, years later, I would join the adjunct faculty teaching Supreme Court litigation) and there's no better place to go for practicing law in Texas.

But there was another school that I had my sights set on. I knew that if I had the grades and test scores to get into Harvard Law School, I couldn't say no. The novelist Scott Turow had captured the intensity of the school's first-year experience in *One L*, and the bestselling novel *The Paper Chase* had become a movie and then a television show. It was hard to imagine anything that might mean more to my immigrant father than for his son to attend Harvard Law School.

Besides its iconic status, Harvard promised some of the best professors and students imaginable. I wanted to learn from them all, inside and outside the classroom. I was excited to imagine sitting in a classroom, debating constitutional law with a professor who literally wrote the (case) book on the subject and with a student who might one day be a Supreme Court justice. Sure, many of them would be liberal. But that was no deterrent. I wanted to test my beliefs against the best the other side had to offer, and I couldn't think of a better place to do so.

To my great relief, I got in. And, to make it even better, so did David. He and I would be going to Harvard together, roommates once again.

CHAPTER 4

May It Please the Court

Tyler, Texas, was named for an obscure former president of the United States. It was a community of some 85,000, about a hundred miles east of Dallas.

The man driving through the city's streets in the seven-year-old Mercedes on April 19, 1994, was John Luttig.[1] He was heading home from Dallas with his wife, Bobbie, seated in the car beside him.

In their rearview mirror, the couple could catch sight of a car following behind them. Its occupants were a group of teenagers and its driver was a man with angry eyes and a .45-caliber pistol at his side. Napoleon Beazley told his friends that night that he wanted to see what it felt like to kill somebody.

The Luttigs pulled into their garage, hoping that the pursuing car would drive away. Instead, Beazley's car swerved in right behind them. Beazley exited his car and stripped off his shirt, carrying the pistol in his hand. Another passenger, Donald Coleman, followed behind with a sawed-off shotgun.

Before the 63-year-old John Luttig could take any action, Beazley ran toward his Mercedes and fired one round. The bullet hit Luttig on the side of his head. Then Beazley's eyes turned toward the woman in the passenger's seat.[2]

As her husband lay gravely wounded, a horrified Bobbie Luttig pushed herself out of the passenger's side of the Mercedes in an attempt to flee. Beazley fired a shot at her, and thankfully missed. Still, she fell to the ground and rolled under the vehicle. As she later put it, "I was wondering what the bullet would feel like if it went through my back. I was wondering what it would—how it would feel to die."

Satisfied, Napoleon returned to his first victim, raised his gun, and fired a fatal shot.

Bobbie Luttig heard her husband cry out as he died. "I had never heard a sound like it," she said, "never heard the horror that was in it, in his voice."[3]

Then, as Beazley stood in the victim's blood, he looked for the keys to the Mercedes.

His attention turned again to the other victim of the attack. "Is she dead?" he asked.

Still carrying the sawed-off shotgun, Donald replied that, no, she was still moving.

"Shoot the bitch!" Beazley ordered.

Donald at first refused. When Napoleon insisted, Donald replied that he was wrong. She already was dead, he lied.

Beazley then took the Mercedes and pulled it out of the driveway. Later commenting on the carjacking and murder, this evil man was heard to comment that the experience "was a trip."

The brutal and senseless murder of John Luttig electrified the city of Tyler and soon received national attention. The Luttigs' son, Michael, was one of the most well-respected judges on the federal court, a close friend to a number of justices on the U.S. Supreme Court. And he happened to be my boss.

The murder of John Luttig occurred shortly after I had been hired as a clerk at the U.S. Court of Appeals for the Fourth Circuit, based in Richmond, Virginia.

A fastidious man with a mop of dark blond hair, Luttig was a young judge. In fact, when he was appointed to the bench at the age of thirty-six by President George H. W. Bush, he was the youngest federal appellate judge in the country.

Judge Luttig was extremely close to his dad. And understandably, the episode was a crushing experience. Every one of us who knew him, perhaps thinking of our own dads, was pained to the core.

While remaining on the bench, Judge Luttig spent much of the next two years following the five separate trials of his father's murderers. He could recite every detail, every witness statement. And he made it his personal mission to ensure that no one, out of a misplaced sympathy for criminal defendants or their own life's circumstances, forgot the evil truth about what had happened to his family on that spring evening.

During sentencing of Napoleon Beazley, Luttig was allowed to give a victim's impact statement to the jury. The whole statement is worth a read, but here is how he described his world in the days after his father's senseless murder:

> It's living in a hotel in your own hometown, blocks away from where you have lived your whole life, because you just can't bear to go back. . . . It's packing up the family home, item by item, memory by memory, as if all of the lives that were there only hours before are no more. . . . It's reading the letters from you, your sister, and your wife, that your dad secreted away in his most private places, unbeknownst to you, realizing that the ones he invariably saved were the ones that just said "thanks" or "I love you." And really understanding

for the first time that *that* truly was all that he ever needed to hear or to receive in return, just as he always told you. . . . It's carefully folding each of your husband's shirts, as you have always done, so that they will be neat when they are given away. . . . It's watching your mother do this, in your own mind begging her to stop. . . . It's cleaning out your dad's sock drawer, his underwear drawer, his ties. . . . It's packing up your dad's office for him, from the family picture to the last pen and pencil. . . . It's reading the brochures in his top drawer about the fishing trip you and he were to take in two months—the trip that your mother had asked you to go on because it meant so much to your dad.

Most moving, Luttig described the year after the murder, going to the graveyard on Thanksgiving so that his dad wouldn't have to spend his first Thanksgiving alone.

It was one of the most powerful things I'd ever read. Even today it fills me with sadness and anger—as well as a great deal of pride for a man who suffered such a devastating loss but still carried on the duties of his office with integrity, honor, and a commitment to the truth.

———

I was standing behind the Chief Justice of the United States and sitting right next to him was the Court's first female justice, Sandra Day O'Connor. We were in front of a large computer screen gazing at explicit, hard-core pornography. As we examined the screen before us, I remember very distinctly what the sixty-five-year-old O'Connor said.

Before I finish that story, maybe I should start from the top.

As a kid, I didn't start out wanting to be a lawyer. No one in my immediate family had a law degree, and we didn't really know any lawyers. But after studying the Constitution in high school, I was hooked.

In college I spent several summers working at a large Texas law firm to make money. At first I was making copies and performing clerical duties, but over time I pestered the lawyers to let me do some substantive legal research, so much so that they finally gave in. (I am pretty confident they didn't rely on my work for much of anything.)

As I think back, one of the things that really attracted me to the law was a 1979 book by Bob Woodward and Scott Armstrong, *The Brethren*. It was compiled from interviews with former law clerks who shamelessly depicted what really happened inside the Supreme Court and chronicled the justices' deliberations on major Court decisions.* The book was a serious violation of the trust that the individual justices traditionally place with their law clerks. Whether it was taboo or not, as a young teen I found the book fascinating.

I try hard to be precise in the things I say. And there was something about the logic of the law, the sense of meting out justice, the respect for precedent and legal canons, that spoke to me. Upholding and respecting the law is the foundation of the American system of justice, which in turn is the very foundation of our democracy. In high school, I decided that being a law clerk at the Supreme Court, working alongside nine of the most revered judges in the land, would be one of the coolest jobs in the world.

Upon entering Harvard Law School, David and I shared a two-bedroom suite at Walter Hastings Hall. The redbrick building, built in 1888, presented a picturesque law school setting—it had bay

---

* According to Woodward, one of the sitting justices, Potter Stewart, had been a key source for the book as well.

windows, wood-paneled rooms, and built-in fireplaces. Everything about it suggested history.

My first class in law school was on the subject of property law, and was taught by Professor Charles Donahue. He was a liberal academic straight out of central casting—he had a long beard and wore Birkenstocks. He looked something like a lumberjack who'd found his way to Berkeley. And he was, in fact, a wonderful teacher.

On the first day of class, he walked to the blackboard and wrote a long Latin phrase in chalk:

Sibilisi ergo, fortibuses in ero. Nobili demis trux. Sewatis enim? Cowsendux.

Then he turned back to a room filled with law students anxious to impress and please their professor on the first day of classes.

"Does anyone know what this phrase means?" he asked. "Can anyone guess?"

I certainly didn't. In fact, the phrase meant nothing. It was gibberish—but if you read it aloud in English it was "See, Billy, see 'er go. Forty buses in a row. No Billy, dem is trucks. See what is in 'em? Cows and ducks." Just because something looks like it has a profound meaning doesn't mean it does. From that, Professor Donohue offered us his first lesson: "Don't check your common sense at the door."

Probably the most well-known professor I encountered was Alan Dershowitz. Dershowitz is an extraordinarily accomplished criminal law professor and attorney who has represented such famous figures as Michael Milken, Patty Hearst, Leona Helmsley, and Mike Tyson. In 1990 he had been portrayed by actor Ron Silver in the feature film *Reversal of Fortune*, about his successful efforts to overturn the conviction of British socialite Claus von Bülow for the attempted murder of his wife in Newport, Rhode Island.

I was at Harvard when Dershowitz joined the so-called dream

team in the defense of the accused double murderer O. J. Simpson. The trial did not reflect, probably by anyone's standard, how a respectful, professional legal proceeding should transpire. It was a tawdry media circus that exploited, for the media's entertainment, the brutal murders of two innocent human beings. I do, however, recall Dershowitz assigning us to read an article about the case that he had written for *Penthouse* magazine, which, as it happened, was not typical required reading at one of the nation's most respected law schools. But the media-savvy Dershowitz pointed out, "You know what? The readership of *Penthouse* is five million people. If I'm trying to get something in front of would-be jurors, that's a pretty good outlet in which to do it."

Along with his brilliant mind, "Dersh" had the heart of an advocate. He was in his mid-fifties when we met, with wavy dark hair and inquisitive eyes behind wire-rimmed glasses. He loved argument and debate, and unlike many Ivy League academics I've encountered, he respected people with differing points of view. He hated nothing more than liberals who reflexively agreed with him but couldn't explain why. To Dersh, the worst thing a student could say to back up his point was "Well, I just feel that . . ."

Dershowitz pounced on such words. "Oh, you *feel*, do you?!" he'd ask. "You're *emoting*?! I thought you were in *law* school." He wanted logic, argument, substance.

In most of my classes I consciously tried to be relatively quiet. I had no desire to be a "gunner"—one of those students who shot his hands up constantly in class and fell in love with the sound of his own voice. To try to prevent that, I deliberately rationed myself to talking just once a week in my classes. That practice in itself turned out to be an interesting exercise, with a positive effect. Restricting what you say forces you to be selective, which tends to make your comments smarter and your points sharper.

But I broke my self-discipline—repeatedly—in Dershowitz's criminal law class. He frequently harangued the opinions of conservative justices on the Supreme Court—with his pointed verbal jibes aimed particularly at two of the justices whom I admired most, Antonin Scalia and Clarence Thomas. He did this so often that it ticked me off and invariably prompted me to raise my hand: "Now, hold on a second, professor."

In fact, many times our arguments would continue after class, back in his office, where we'd battle back and forth for hours on end.

Dersh genuinely cared about his students. Every year, after our first-semester grades came out, Dershowitz gave the first-year students, or 1Ls, a pep talk. "Every student here was a top student in elementary school, a top student in high school, and a top student in college," he'd begin. Then he'd add, with just a slight grin, "Yet even at Harvard, fifty percent of our students . . . end up in the bottom half of the class."

I had no intention of being among that group. Wanting to be a Supreme Court clerk, I had researched the qualifications I needed. Most obvious, I needed to have stellar grades. One of the first things I did was form a study group, which consisted of three students: me, my roommate David, and Jeff Hinck, an economics major from Northwestern University whose native brilliance and midwestern common sense made him one of my closest friends in law school. Our study group met in our dorm room several times a week throughout the first year of law school.

I undertook my mission with considerable focus. At the time, I was dating a woman who was getting her economics Ph.D. from MIT. Whenever I was studying at her apartment, I would bring six different colors of highlighters to annotate my casebooks. Seeing me with a handful of highlighters, she and her friends could only laugh. And when you are being mocked by MIT graduate students for geeking out, you really have a problem.

I did take time for at least one diversion, one that inadvertently taught me a valuable lesson. I joined the drama society and was cast as Revered Paris in *The Crucible*, Arthur Miller's powerful play about the Salem witch trials. The first night of the play went very well and to celebrate we had a raucous—very raucous—cast party. Being a stupid twenty-two-year-old, I had way too much to drink, not giving a thought to the repercussions. During the next day's performance, I was still sick—so horribly sick that, in the middle of the performance, I walked off the stage and curled into a ball behind it. My startled fellow cast members were left to ad-lib the rest of the scene without me. As it happens, my sad display was captured on tape, a copy of which found its way into the hands of a reporter from the *Boston Globe*. As I told the *Globe* reporter who asked me about this years later, young people are not known for their wisdom or their discretion, and I was no exception.

While I was learning life lessons in law school, I also tried out for the law review. Law reviews are student-edited legal journals, and just about every Supreme Court clerk has been on law review. At Harvard, people still talked about the first black president of that prestigious legal journal, named Barack Obama, who'd graduated just one year before I arrived at law school. To get on the law review, I participated in what amounted to an eight-day legal writing and editing competition. I filled one of forty slots that year—thirty-two of which were based on grades or writing. The remaining eight were based on affirmative action.

In my second year, I decided to run for president of the law review. I wanted the position, of course, because it provided as close to a lock on a Supreme Court clerkship as one might get. I soon learned that I had zero chance of winning.

Election of the law review president was a ruthless process, one worthy of study by Machiavelli. The election was not necessarily for the brightest among us, or the most accomplished, or the most articulate. None of us wanted someone like that getting the job and thus increasing his or her odds of getting a Supreme Court clerkship at our expense. Rather, the way you got elected was to demonstrate that you posed as little threat to other law review colleagues as possible. Those without any interest in a clerkship or those certain to get one anyway were thus immediate front-runners.

You could not be outspoken. Eleven people ran for the position that year—and those of us who were the most opinionated were quickly winnowed out. I was among them. As a child of someone who fled Cuba, I was not amused by the trendy Marxist philosophy espoused by some of my colleagues in the editors' lounge. I also found myself in heated exchanges over the law review's affirmative action policy, with which I strongly disagreed.

Earlier that year, a group of conservatives on the law review had decided to challenge the affirmative action policy. I wasn't terribly in favor of doing so—I could count the votes, and the liberals had an easy majority—but the conservatives went forward anyway. All eighty student editors gathered in a classroom for a robust debate, and initially I stayed quiet. Then one fellow turned to all of us and said, "If we abolish affirmative action, the *Harvard Law Review* will be nothing but rich white men."

This was said with total sincerity, and it is sadly a view shared by many well-meaning liberals who think they are doing the right thing. But it also proved in a single sentence what was wrong with affirmative action.

Finally, I raised my hand. "You know what," I began, "that last comment perfectly embodies how insidious affirmative action is." I pointed out that the comment, on its face, implied that not a single person in the room who was not an Anglo white male deserved to be there. That we couldn't make it on merit, that we couldn't rise to the top without the help of our betters, fueled by their liberal guilt.

The comment was even more revealing, I observed, because at the time the law review *did not have affirmative action for women*; the affirmative action policy was purely on racial and ethnic lines. And yet this supercilious liberal had suggested that no women would make the law review if selections were based purely on merit.* What nonsense.

The argument was heated and personal. And then the votes came down to preserve the affirmative action policy, just as was apparent at the outset.

Likewise, I quickly lost the election for president of the law review. Later that evening, I was instead elected to be one of four "primary editors," essentially the lead student editors for academic articles. But even if I had had the opportunity to do it over again, to be noncontroversial and universally amiable, I couldn't have done it. Those qualities are simply not who I am.

When David Panton came back to the law school two years later— he had left Harvard to attend Oxford on a Rhodes Scholarship—I helped with his successful campaign for the law review presidency.

"Say nothing controversial on any topic," I advised him. "Go out of your way to be charming and nonthreatening to every person there." This, I knew, would be easy for my friend. He also made it clear to his law review colleagues that he was not interested in a Supreme Court clerkship. He wanted to go into business in Jamaica. When David was elected as the second black president of the *Har-*

---

* Several times previously, the law review had voted on whether to extend affirmative action to women. Each time it was voted down, with many of the liberal women opposing any expansion; they knew full well that doing so would devalue the credential they had worked so hard to earn.

vard Law Review, he received a congratulatory call from the first: Barack Obama.

During a summer job at a law firm, I met a lawyer who had clerked for Chief Justice William Rehnquist. He told me that even if I had high grades and made it onto law review, so would a lot of other students applying for the positions. I needed to realize that I was dealing with a small universe of people on the nation's high court, he said. The justices were nine human beings making different hiring decisions based on their own personal criteria. They deferred heavily to the recommendations of people they personally trusted—such as professors at the law schools they respected. The justices would call them up and say things like, "I've got applications from a number of your top students. Which one should I really go with?" Those recommendations, the lawyer told me, were critical.

Thus many students who wanted clerkships took jobs as research assistants for connected law professors—those who tended to have relationships with various Supreme Court justices. The arrangement was understood by all the participants. The professors received relatively inexpensive legal work from bright students working their tails off. And the successful students would receive recommendations.

I worked for three different professors: David Shapiro, Daniel Meltzer, and Charles Fried. A brilliant man and expert on civil procedure, Shapiro was one of the few living summa cum laude graduates of Harvard Law School. Meltzer would eventually serve as deputy White House counsel under President Obama. Fried had been a solicitor general for Ronald Reagan and was the lone outspoken Republican on the Harvard faculty. I worked the most for Fried. One time, when I was helping him with a case he had before the Supreme Court, I put in 104 hours of work in a single week. As a result, I missed several of my classes that week, including his class.

Fried gently admonished me, "Do I need to scale back your work so you can attend classes?" I made sure not to miss his class again.

As the most influential conservative at Harvard, Professor Fried's recommendation carried a great deal of weight with justices like Scalia, Thomas, and Rehnquist. But when I prepared my application I also included a recommendation from someone on the opposite end of the spectrum—Alan Dershowitz.

It is a curious fact that most of the more conservative justices on the Court (Rehnquist, Scalia, Anthony Kennedy, O'Connor) tended to hire at least one liberal clerk. The liberals on the Court (John Paul Stevens, David Souter, Ruth Bader Ginsburg, Stephen Breyer) almost never hired a conservative. That explained a lot—contrary to the conventional wisdom, liberals in my experience tend to be far less open-minded and welcoming of diverse opinions than conservatives.

———

Once I had my recommendations in order, the next step toward the Court was trying to secure a one-year clerkship with a federal appellate judge.

In my perhaps excessively methodical way, I had mapped out a list of all the potential "feeder" judges to the Court—those appellate judges whose clerks tended to most frequently obtain clerkships on the U.S. Supreme Court. Because he had only been on the bench for three years, Michael Luttig hadn't even appeared on my list. That was a significant omission on my part.

Luttig has had a remarkable legal career. He was Antonin Scalia's very first law clerk when Scalia served on the U.S. Court of Appeals for the District of Columbia Circuit. Luttig worked in the Reagan White House, and he worked at the Supreme Court for four years as then–chief justice Warren Burger's assistant and then law clerk.

He and Burger were so close that he was executor of Burger's will when he passed away. He knew everybody at the Court, and they trusted him.

Thus Luttig, in a short amount of time, had become one of the top feeder judges in the country. I learned that of the nine clerks he'd had since he took a seat on the bench, seven had gone on to clerk at the high court. I quickly faxed an application over to his office, interviewed with him in Richmond, and we hit it off immediately. He and I became very close; indeed, for many years he was like a father to me.

When you work for Michael Luttig, you put in eighteen- to twenty-hour days. He was a perfectionist, and especially hard on himself. He could go through fifty or sixty drafts of opinions, which he typed himself.

He took personal responsibility for training his clerks to be excellent lawyers. He knew every clerk well and cared about their personal and professional development. When he spoke on the phone with other judges, Luttig would often have us in the room with him so we could learn how jurists communicated with each other. He had set up three computer monitors on the table adjoining his desk so clerks could sit beside him and watch him draft opinions.

Luttig is an immensely meticulous man. He wrote with only a certain black felt-tip pen. A paper clip on material would have to have the bigger side on top of the page and the pointy end on the inside. If the paper clip wasn't right, he'd send the document back.

He insisted that clerks offer clear recommendations on how he should rule on cases, especially the hard ones where a case could be made for either side. This was an important lesson: The time your judgment matters most is often when the decision is the hardest. And he used gentle ridicule to teach; so if a clerk said, "This case is a really close call, so it's up to you," he'd reply, "Oh, so you only have an opinion on the easy ones?" And so you'd go back and think harder about the case.

He valued concision. Our bench memos—summaries of a case's arguments, relevant case law, and recommendations—needed to be ten pages; if one was longer, he told us, we hadn't really figured out what the case was about. And he had two favorite sayings. The first was somewhat facetious: "Never muck up a good story with the truth." We laughed uproariously with Judge Luttig, a master storyteller. The second saying reflected great wisdom, especially in regards to the bloated federal government: "Never attribute to malice what can be explained with incompetence."

———

Just weeks before I started my clerkship with Judge Luttig, I was working at the Houston law firm Baker Botts, fresh after graduating from law school. The most notable occurrence that summer of 1995 was that the Houston Rockets were going to the NBA Finals for the second year in a row. Major law firms love to wine and dine their summer associates in the hope that they will come to work for them after graduation, when the firms will work them around the clock. And it was my good fortune that Baker Botts was outside counsel for the Rockets, which meant they had an abundant supply of tickets to the playoffs.

I'd worked at Baker Botts in 1994 as well, and so was thrilled to get to attend game seven of the finals, in which we beat the New York Knicks to win the first ever major-league championship for my hometown. Sports can really bring people together, and the Bayou City rejoiced; I saw grown men in tears. People in business suits hugged homeless people in the streets.

That next year, the Rockets were again going to the finals. In fact they were headed to game four of the series, getting ready to sweep the Orlando Magic, and there was no place on earth I more wanted to be. Well, with one exception.

The day before the game, I received a phone call from the chambers of the Chief Justice of the United States.

"The Chief would like you to come and meet with him," said the friendly voice on the other line. "When are you available?"

"I'm available whenever the Chief Justice would like," I replied quickly.

"Can you come in tomorrow?"

Tomorrow. The day of the big game. The game I'd waited for. "Yes," I replied. "Of course."

Some justices liked to have lengthy give-and-takes with potential clerks—Scalia being the most notorious. Chief Justice Rehnquist preferred shorter encounters, hardly ever lasting more than twenty minutes. That, in the Chief's view, was sufficient time to decide on a person. He knew that all of the aspiring clerks applying to his chambers had strong academic credentials. What he really wanted to see from in-person meetings, I figured, was whether he liked you, whether it would be bearable to work with you every day for the course of a year.

William Hubbs Rehnquist was born in Milwaukee, Wisconsin, and was very much a midwesterner. He was polite, low-key, and modest. An enlisted man during World War II, he was a weather observer in North Africa, leading to a lifelong fascination with meteorology. Like Scalia, he had a sharp wit, but he tended to wield humor more gently. Justice Lewis Powell, for example, was a colonel in World War II. "You know," he once quipped to the Chief, "I outrank you." Without missing a beat, Rehnquist replied, "Not anymore."

As an attorney in Phoenix in the 1960s, Rehnquist had worked for the Barry Goldwater presidential campaign. He had been a law clerk to Justice Robert Jackson, one of the finest justices in the history of the Court. Rehnquist, who went on to work in the Nixon administration as the head of the Office of Legal Counsel, liked to joke about his former boss Nixon, who, on one of the White House tapes discussing Supreme Court appointments, was overheard saying, "What about that clown, Renchburg?"

For many years on the Court, Rehnquist was known as the "Lone Ranger," since in the 1970s he was often the only dissenter against the liberal-minded Warren Court. But slowly, over time, many of Rehnquist's dissents became the law of the land as he shepherded the Court, and the law, back to its foundations. There are few clearer arcs than the trajectory from Rehnquist's dissents in the 1970s and '80s to the majority opinions of the 1990s and 2000s.

By the time I arrived to interview with the Chief, he had already begun to usher in the so-called federalist revolution, which restored some of the historical deference toward the authority of sovereign states. As the Chief Justice wrote in the case *United States v. Lopez*, "We start with first principles. The Constitution creates a Federal Government of enumerated powers, in James Madison's word, which 'ensure[s] [the] protection of our fundamental liberties.'" That transformation—along with the revival of religious liberty and the restoration of balance in criminal justice law—had earned Rehnquist the designation by President Clinton's acting solicitor general Walter Dellinger as "among the three most influential Chief Justices in history."

I soon learned why Rehnquist spent so little time interviewing his law clerks. He was one of the most brilliant human beings on the planet. He could have done the job with his eyes closed and without a single assistant.

I met the Chief in his spacious chambers at the back of the Supreme Court Building, right behind the courtroom. A tall man with thinning black hair, long sideburns, and oversize glasses, he ambled up and offered me a chair. The Chief wore big wide ties with floral patterns and Hush Puppies.

With his first question, I knew that my recommendation strategy had paid off. "You know, I've got two recommendations here from Charles Fried and Alan Dershowitz," he said, in his deep, gentle voice. Rehnquist knew as well as anyone how diametrically opposed the two men were. He had a slight grin. "So, I wonder, how on earth is that possible?"

"Maybe one of them was confused," I said. He laughed.

I told him a little about my background. When I described my father's background, fighting in the Cuban Revolution, it piqued the Chief's interest. Afterward, when I recounted the interview to my dad, I kidded my father by telling him that I had described him as a communist guerrilla for Castro. My dad was horrified.

"I was never a communist!" he said indignantly. He was flabbergasted that the Chief Justice of the United States might have this mistaken impression.

As we continued talking, Rehnquist asked a question that seemed of great importance to him.

"Would you be willing to play tennis with me and my other clerks?"

He was, as I had learned, a devotee of the game. In fact, many of the clerks he tended to favor were skilled tennis players or even all-American athletes.

"Sure, sounds like fun," I replied. I wanted the job badly, though I had to be honest. "I should tell you I'm not very good."

Rehnquist laughed at what he apparently took for false modesty. What he didn't know, but would soon learn, was that "not very good" was a boast bordering on hyperbole.

Walking out of the interview, I decided to take a gamble with a joke. "You know, being here today is actually quite bittersweet for me," I told him.

"Really?" he said, somewhat taken aback. "What do you mean?"

"Well, today is game four of the NBA Finals. And my Houston Rockets are about to sweep back-to-back championships. And I gave up my ticket to come talk to you."

I knew from reading about him that Rehnquist was a sports fan. And as I had hoped, he cracked up laughing.

Then he said, "Well, I think you made the right decision."

To this day I am convinced that the reason I got the clerkship I'd coveted for years was that on the spur of the moment I made the Chief Justice laugh.

———

Over the course of my time at the Supreme Court, I came to realize that the clerks tended to reflect the characteristics of the justices they served.

David Souter had been appointed by George H. W. Bush but fairly quickly became one of the more liberal justices. Raised in rural New Hampshire, he lived a simple, spartan life. When he hosted the clerks for lunch, he explained that each day he would have a bowl of yogurt. On the weekends he would have a bowl of yogurt, but with fruit.

I remember thinking, "He prefers it with fruit." And it was interesting to me that he chose to deny himself that pleasure during the week.

Justice Stephen Breyer, a 1994 Clinton appointee, can be delightfully charming. He did have a habit of speaking quite loudly about pending cases—in places like public restaurants. On occasion that would cause his clerks considerable consternation.

No member of the Court is more reviled on the left than Clarence Thomas. At the same time, no member of the Court is more beloved by the Court's janitors, guards, and support staff members, with whom he connects on a real, personal level.

I admire Justice Thomas. He rose from abject poverty—having grown up penniless in Pinpoint, Georgia—to the pinnacle of the law. A brilliant, scholarly jurist, he has been unfairly maligned by liberal academics and journalists, enduring condescending insults that never would have been directed at a liberal or at a white con-servative. (Justice Scalia, every bit as conservative, has never been depicted on a magazine cover as an Uncle Tom, licking another jus-tice's boots.)

With his welcoming demeanor and deep, hearty laugh—imagine Santa Claus bellowing "ho, ho, ho"—Clarence Thomas has carried out dozens of acts of kindness on the Court, the kind never reported by the mainstream media. An illustrative story involved one of my co-clerks, Rick Garnett, who had worked the previous year as a clerk in Little Rock, Arkansas. There he and his wife had befriended and tutored a young African-American boy named Carlos. The boy had never left Arkansas before, but Rick and his wife paid to fly him up to spend some time in D.C. Rick emailed all nine chambers at the Court, saying that this young boy would be in town, and asking if any of the justices would be willing to meet with him. Two offices responded—those of Justices Ruth Bader Ginsburg and Clarence Thomas. Ginsburg is an incredibly talented lawyer and jurist, and it was very kind of her to meet with Carlos, but her prim demeanor is that of a legal librarian, and so it was difficult for her and the young boy from Arkansas to connect. Clarence Thomas understood the world that Carlos had come from.

At the end of their two-hour conversation, Carlos observed that Justice Thomas was a Dallas Cowboys fan. (Thomas had a framed picture of himself with quarterback Troy Aikman in his office.) The

kid was very impressed—that was way cooler than the Supreme
Court—and Thomas noticed. So Thomas rose from his chair,
walked to his desk, and showed the boy a Super Bowl ticket, encased
in Lucite, and signed by Cowboys running back Emmitt Smith. He
handed the ticket to the young man.

"I'm going to give you this," Thomas said. "But I want you to
promise me that you will get A's in school next year."

The young man, astonished and wide-eyed, nodded in agreement.

Justice Scalia's clerks, like the justice himself, tended to have
an edge. They were wickedly smart, engaging, and had no problem wielding sharp elbows when warranted. Scalia himself had a
mischievous sense of humor. One famous Scalia story—there are
many—occurred during the 1980s, when Reagan was president and
considering appointments to the Court. Everyone knew that two of
the stars on the conservative side, and thus possible nominees, were
Robert Bork and Scalia, both on the D.C. Circuit. So one day Scalia
was walking in a parking garage at the appellate court when two
U.S. marshals stopped him. "Sorry, sir," one of them said. "We're
holding this elevator for the attorney general of the United States."

Scalia pushed past them, entered the elevator, and pressed a button. As the doors closed, Scalia shouted out, "You tell Ed Meese that
Bob Bork doesn't wait for anyone!" And, as it happens, Scalia was
nominated to the Court by President Reagan in 1986.

Like his polar opposite Dershowitz, Scalia enjoyed having analytical arguments with the other side, as it tended to sharpen his
own view. For his clerks, Scalia usually hired three conservatives
and one liberal. But this practice sometimes had an unfortunate side
effect. A number of Scalia's liberal clerks would go on to become
law school professors and would publicly criticize the conservatives

on the Court. But because they had clerked for Scalia, some media portrayals would leave viewers with the false impression that these liberal critics were conservatives denouncing their own. It was less than ideal.

As Chief Justice, William Rehnquist had the statutory ability to hire five clerks instead of the customary four. But he usually hired just three. Rehnquist was an "if it ain't broke, don't fix it" kind of guy. Back when he began serving on the Court, three was all a justice needed. And he saw no reason to change that.

There was another, apocryphal reason why Rehnquist kept three clerks, which was that it was perfect for doubles tennis. The Chief had not been kidding when he first asked me if I'd play. In fact, it was a required part of the job to play tennis with him once a week— Thursdays, without fail, at 11 a.m.

Before then, I'd probably played the game two or three times in my life. None of my friends played tennis; I hadn't exactly hung around country clubs in my youth. So for months before I started my clerkship with the Chief, I practiced tennis. I took lessons, even hired a coach.

I was still dreadful, despite all my practice. And my skills would soon be put to the test.

The chief traditionally took the best tennis player as his partner and in my year that was a clerk who had played JV tennis for Yale. In our first game, they won all three sets, 6–0, 6–0, 6–0. The second week the result was exactly the same. I was painfully embarrassed.

Finally, the Chief threw up his hands. He'd had enough. I'm sorry to say that I have the ignominious distinction of being the only law clerk who was so bad at tennis that the Chief gave up the best tennis player as his partner and paired him up with me just to try to even things out.

For our next games, I managed not to screw up too much. I stayed in a small part of the court and let my co-clerk dominate the rest. He was a good enough player that we ended up winning, but the games were close. So for that year—and probably only that year—the Chief Justice had a losing record against his clerks, which did not exactly please him. As unfailingly kind and polite as he was, William Rehnquist was a very competitive man.

He wagered on anything, everything—the amount of snowfall that would come to Washington, the outcome of an election or sports game. The bet was always the same—one dollar—and he almost always won.

Once, when he and I were teammates playing croquet, another game he loved and I wasn't terribly good at, he looked at me with a discouraged glance. "Ted, you do know the point of the game is to win, don't you?"

My failings on the playing fields not withstanding, the Chief Justice was a wonderful boss. All nine of the justices on the Court at that time—Rehnquist, Scalia, Thomas, O'Connor, Kennedy, Stevens, Souter, Ginsburg, and Breyer—were extremely intelligent. But even in that room Rehnquist stood out. He had been first in his class at Stanford Law and he had a photographic memory, unlike any I'd ever encountered.

My first impression of him proved right. He really didn't need law clerks. He was so damned smart.

Most of the justices had law clerks prepare long bench memos on cases. Rehnquist didn't need that. He just wanted three-page summaries of the facts. He knew all the rest already.

The Court heard about eighty cases during the term I clerked, and each of the Chief's clerks was responsible for knowing about one-third of them. It was a lot to juggle in your head without notes. And the way the Chief Justice liked to prepare for oral arguments could be disconcerting. He could come by our desks at any time to talk about a case.

"Ted," he might say, "are you ready to discuss *Smith vs. Jones?*"

We knew what came next. I'd say, "Yes, Chief," and get up and go outside for a walk with him. Without any notes at my disposal we'd discuss that case's merits.

As we strolled down First Street across from the U.S. Capitol, he'd say things like, "So, Ted, what did you think of the argument in footnote seventeen of the petitioner's brief? I didn't find it very persuasive."

"Uh, I agree, Chief," I'd respond, struggling to remember the footnote he was citing.

We would do laps around the Court. A kindly looking gentleman, often wearing a cap, Rehnquist was frequently stopped on the street by tourists asking him, a passerby, to take photographs of their families standing in front of the Court or the Capitol. He was so down-to-earth, so approachable, that he could be as comfortable with a plumber as he could a poet. And he was hardly ever recognized.

Each time he was stopped to take a photo, the Chief would smile and say gamely, "Sure."

To this day, hundreds of people have had their picture taken in front of the Court by the Chief Justice of the United States and never knew it. That made him chuckle.

He was not a man of airs. For lunch, he would usually order the same thing—a simple menu of a cheeseburger and a Miller Lite, or as he called it, "a Miller's Lite." And he'd smoke a single cigarette. His townhome in Arlington, Virginia, was modest; there were no signs of pretense or grandeur. Sometimes he had the clerks over to play charades. My favorite memory of him remains the time he grabbed a slip of paper, fell to the ground, and lay on his belly pantomiming firing a rifle. "Pow! Pow!" he called out. (No one told the Chief that you didn't talk in charades.) His charade was "All Quiet on the Western Front."

Rehnquist was a study in contrast from Justice Scalia. Scalia, in private, is every bit as he seems in public—volcanic, passionate, temperamental, brilliant, animated. Anyone who spots him having lunch during the month of June, when major decisions tend to come down, can tell in an instant by his demeanor if his side is winning or losing.

William Rehnquist was very different. He was a stoic, comfortable in his beliefs and his values, and entirely at peace if the Court ruled a different way. He had faith that truth would prevail and that eventually the Court would come around.

One sobering component of being a Supreme Court clerk literally involves life-and-death decisions. A great many states set executions late at night, often at midnight. Just before that deadline, under the current system, defense lawyers tend to file a flurry of last-minute appeals to the Supreme Court for emergency stays of execution.

A clerk from each justice's office would stay up late at night waiting for appeals to come in. Sometimes, as little as an hour before an execution, a fifty- or hundred-page fax would come into the Court (it was the age of faxes), laying out the case for why a defendant's trial was unfair and he should have a new hearing, or raising some other legal point justifying a stay.

If the appeal was in a circuit assigned to your justice, it was your responsibility to read through the entire petition as quickly as possible. Then the clerk would call his boss, who was at home and probably asleep (although they knew the call was coming). In my case, the conversation would go something like this:

"Mr. Chief Justice, there is a last-minute appeal in the X case. They've raised six claims here to justify a stay." Then I'd describe the claims over the phone. I'd give my recommendation, and then the Chief Justice would decide how he wanted to vote. In most cases, justices vote against a last-minute stay of execution. I'd then write up a draft memo for the other justices' offices, summarizing the appeal and outlining why the Chief had voted to resolve the claims as he did.

I would also make a point of doing something the liberal clerks who opposed capital punishment rarely did—simply describing the brutal nature of the crime for which the defendant had been convicted. The appeal would go to the full Court for a vote. Each of the other eight clerks would call their justices at home, wake them up, and then the justices would each vote on the appeal that night.

The current system is indisputably being gamed by the defense lawyers in the hopes that the clerks will get these voluminous appeals filed at the last minute, throw up their hands, and the Court will issue a stay just to have time to sort through it. It's a cynical strategy—in my view, a dangerous one—which results in claims being adjudicated by sleepy justices at midnight with little time for adequate review.

I still don't understand why the justices tolerate this gamesmanship, which diminishes respect for the rule of law. It would be a simple matter for the Court to issue rules saying that all applications for stays must be filed at least one week before the execution date. Or, even a single justice could simply announce publicly that he or she would not vote to stay any execution if the stay application were filed less than a week earlier. That would allow, fair, careful, reasoned consideration of the legal claims, not haphazard skimming at midnight. The rule could exclude claims of actual innocence—which could be filed at any time whatsoever—but the vast majority of capital defendants make no claims of innocence.

But that's not how it works. So, late at night, after the Court had voted to deny a stay of execution, it would fall to us clerks to draft the order of the Court.

The lead clerk that night would sign it, with your justice's initials and your own—in my case "WHR/tc"—and then you'd place it on the fax machine and send it to the prison that was about to carry out the execution. Every time, it was a sober responsibility to sign those documents and fax them off.

I believe in the death penalty. It's a matter of justice. Life is precious, a gift from God, and when a person willfully and deliberately murders another, then the ultimate penalty is justified. Working on capital cases, as I have over the years, you see the face of evil—people who have abused, tortured, and murdered the innocent without a hint of remorse. (That's why the liberal clerks would typically omit the facts; it was harder to jump on the moral high horse in defense of a depraved killer.)

On the death penalty or any other major issue before the Court, Rehnquist hardly ever lobbied other justices, though sometimes we would beg him to. It would have been particularly helpful if he had tried to persuade his old friend and law school classmate, Sandra Day O'Connor, to vote his way. Rehnquist and O'Connor had briefly dated, in law school, and they had been friends for more than five decades. She also tended to be the pivotal swing vote on any number of decisions.

But Rehnquist usually refused such entreaties. He was very respectful of the other justices' points of view. "No, no," he'd say. "She can decide this for herself."

This brings me back, finally, to the pornography story. The year I was a clerk on the Court, the Internet was nascent technology. The Court was considering one of the first cases challenging the constitutionality of a law passed by Congress to regulate Internet pornography.

Most of the justices were in their sixties or older. Few knew much of anything about the Internet. So the librarians of the Court designed a tutorial for them. They set up sessions for two justices at a time and their clerks. As it happened, our Rehnquist group was paired with Justice O'Connor.

In a small room gathered the Chief, Justice O'Connor, and their respective law clerks. The librarians' purpose was to demonstrate to the justices how easy it was to find porn on the Internet.

I remember standing behind the computer, watching the librarian go to a search engine, turn off the filters, and type in the word *cantaloupe*, though misspelling it slightly. After she pressed "return," a slew of hard-core, explicit images showed up on screen.

Here I was as a twenty-six-year-old man looking at explicit porn with Justice Sandra Day O'Connor who was standing alongside the colleague she had once dated in law school. As we watched these graphic pictures fill our screens, wide-eyed, no one said a word. Except for Justice O'Connor, who lowered her head, squinted slightly, and muttered, "Oh, my."

CHAPTER 5

The Bush Administration

As the Holocaust survivor entered the White House that morning, to receive a medal from the president of the United States, Elie Wiesel's thoughts were troubled. He had considered boycotting the event altogether. And he was in no mood to accept gratitude or congratulations.

Instead, he was on a mission. A daunting one. To confront directly the most powerful man in America and, indeed, the most powerful and respected leader in the world. As he prepared to do so, it likely was not lost on him that there was no other nation in the world that would allow him such an opportunity. For that fact, he was grateful, despite the anger and sadness that filled his heart.

For nearly half an hour, Elie Wiesel and Ronald Reagan met privately. It was by all accounts an emotional meeting in which the acclaimed writer implored the president he admired to reverse his decision to visit a German cemetery, Bitburg, which included the remains of former officers in the SS.

Reagan had not known the cemetery contained Nazi soldiers when he accepted the invitation to visit, but he held firm on his decision to go. He had made a promise to his friend and close ally, West German chancellor Helmut Kohl, who was battling to hold together his majority coalition and all but begged Reagan to hold his ground on the visit in the spirit of "reconciliation" between the German and Jewish people. (Polls showed 72 percent of West Germans wanted Reagan to visit the cemetery.) In their meeting, the president assured Wiesel that he intended to honor the memory of the Holocaust and would make a visit to the Bergen-Belsen concentration camp, where Anne Frank had died.

That was not enough for Wiesel. He could not abide the thought of an American president laying a wreath at a cemetery that included those who had sought the extermination of the Jewish race. Wiesel and his father had been interned at Auschwitz (his father would later die at Buchenwald). The Germans also had rounded up his mother and sisters, one of whom, Sarah, did not survive. Wiesel himself was scheduled for the crematorium at Buchenwald, and would likely have perished as well had the camp not been liberated by American soldiers only weeks later. Privately, Wiesel believed Reagan had been poorly served by his staff, who had put him into a "no-win" situation by allowing Kohl to make the request of Reagan in the first place.

After their meeting, Reagan welcomed Wiesel to the Roosevelt Room and a crowd of reporters and television cameras.

Characteristically eloquent and gracious, Reagan lauded Wiesel as a teacher about the horrors of the Holocaust. The president presented the medal to Wiesel, who handed it to his son.

As the dark-haired Holocaust survivor took to the podium, the room was unusually silent. Reporters and cameraman stared. The president was attentive and quiet. In the background, White House officials lurked nervously.

"You spoke of Jewish children, Mr. President; one million Jewish children perished. If I spent my entire life reciting their names,

I would die before finishing the task. Mr. President, I have seen children—I have seen them being thrown in the flames alive. Words—they die on my lips."

Some reporters' eyes were wet with tears. The president sat grimly, his face silent and sad.

Wiesel saluted Reagan for "being a friend of the Jewish people, for trying to help the oppressed Jews in the Soviet Union." Then, his body turned in Reagan's direction, he added, "But, Mr. President, I wouldn't be the person I am, and you wouldn't respect me for what I am, if I were not to tell you also of the sadness that is in my heart for what happened during the last week. And I am sure that you, too, are sad for the same reasons. What can I do? I belong to a trauma-tized generation. And to us, as to you, symbols are important. And furthermore, following our ancient tradition—and we are speaking about Jewish heritage—our tradition commands us, quote: 'to speak truth to power.'"

When Wiesel had finished, Reagan stood and shook his hand. The controversy did not die that day. Reagan did not relent. But a great moment in democracy had taken place. An American citizen had told the truth to an American president, who had given him every opportunity to do it. The story for me is a reminder of many things—first, that America is a nation that believes so much in free-dom it would even allow such a debate. Second, that even the best leaders sometimes make bad calls.

————

It was almost impossible to believe. After the many sleep-deprived nights on the campaign, after the hours upon hours of strategy ses-sions, memos, policy briefings, and meetings, it was all a waste. The Clinton era would continue. Albert Gore Jr. was about to become the next president of the United States.

My father, Rafael; my Tía Sonia; and my grandmother on the beach in Cuba, circa 1950. It looked idyllic, but reality on the island was anything but.

My father's mugshot, after he was picked up by Batista's thugs. His nose was already broken.

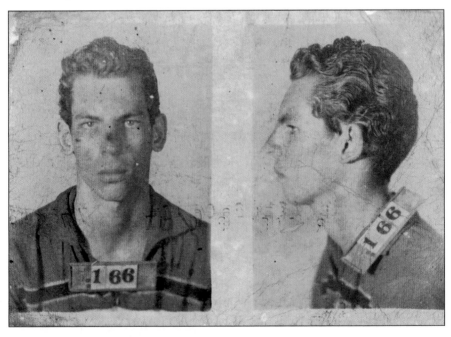

Abuelo and Abuela (my grandparents); my Tía Sonia; and my cousin Bibi after our family had emigrated to the United States—and freedom.

Unless otherwise noted, all photos are courtesy of the author.

My mother, Eleanor, circa 1955. She was the first woman in her family to go to college. She attended Rice University and went on to become a computer programmer for Shell.

My mother with her parents, Edward and Elizabeth Darragh, and her sister, Carol. They were a typical working-class Irish-Italian family from Wilmington, Delaware.

Rafael Cruz married Eleanor Darragh on March 14, 1969.

My parents and I in the 1970s, after we moved back to Houston and they started their own company in the energy business.

My early aspirations to be a Hollywood actor would not come to pass, but my parents always taught me that in America, anything is possible—dream big.

With my half sisters, Miriam and Roxana. Growing up, I was an only child during the school year and a little brother over the summer, when my sisters lived with us.

Reagan's 1980 campaign was my first taste of politics. Here was his last major public address, at the 1992 Republican National Convention. I was near the front row, cheering with hundreds of other college students . . . including my future wife, Heidi. But she and I didn't meet until eight years later.

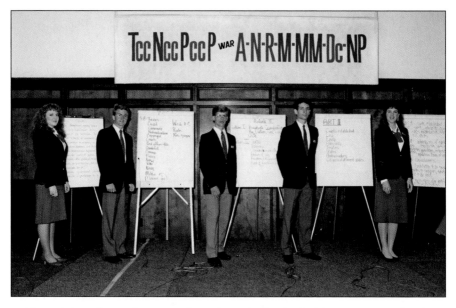

I learned about free-market economics and constitutional principles—not to mention the basics of public speaking—participating in the Free Enterprise Institute and its spin-off, the Constitutional Corroborators. (I'm second from the right.)

At my Princeton graduation with my roommate, debate partner, and best friend, David Panton. We were both headed to Harvard Law School.

Watching the returns the night George W. Bush was elected governor of Texas. I was a busy second-year law student but took the time to savor the victory—and a cigar.

I was deeply fortunate to clerk for William Rehnquist, the chief justice of the Supreme Court. Our first meeting lasted twenty minutes, and his most vexing question was if I'd be willing to join him playing tennis each week.

After my time in Washington during George W. Bush's first term, I returned to Texas to serve as solicitor general. Greg Abbott, my dear friend, who went on to become governor of Texas, was then attorney general—and my boss.
(AP Photo/Harry Cabluck)

As solicitor general, I argued a wide range of cases before the Texas Supreme Court as well as the Supreme Court of the United States.
(AP Photo/Harry Cabluck)

The best thing about my experience on the 2000 Bush-Cheney campaign was meeting a vivacious and brilliant California blonde named Heidi Nelson, who was also on the policy team. As a child, Heidi had traveled with her parents to Africa, where they did missionary work (this picture was taken in Kenya).

Heidi and I were married on May 27, 2001.

Heidi and I were blessed with two little girls, Caroline and Catherine. The three of them are the loves of my life.

Caroline's first day of school in September 2013. Fortunately, both girls look like their mother.

I launched what would most kindly be called an underdog campaign for the United States Senate in 2011. With time, the crowds started to grow as a grassroots army joined our effort.

In October 2011 the conservative *National Review* put me on its cover with the headline "First-Class Cruz." It was a turning point that gave me a national profile and suggested that my candidacy was viable.
(Permission by National Review)

Throughout the campaign, Caroline and Catherine were enthusiastic supporters.

Our first job was to force a runoff in the Republican primary, which would be triggered if no candidate broke 50 percent. I was confident I could win in a head-to-head contest with the establishment candidate, Lieutenant Governor David Dewhurst. We stared at computer screens all night as the good news came in—Heidi was so tense, I thought she would break my wrist. *(AP Photo/Houston Chronicle, Nick de la Torre)*

Once we were in the runoff, our poll numbers swung 25 points in days. But we still had our work cut out for us. When I told Heidi we would need to raise $3 million in three weeks, I thought she would have a heart attack. But we did it, and we won!

After the general election, Heidi and I visited Israel together. It was an amazing spiritual experience, and also an eye-opener to visit the Jewish state and see the amazing success and strength of the Israeli people.

I was sworn in as the junior senator from Texas on January 3, 2013, with Vice President Joe Biden administering the oath.

From my earliest days in the Senate, I seemed to rub some of my senior colleagues the wrong way—especially the leadership of my own party. Senator John McCain, for example, called me a "wacko bird," less than two months after I was sworn in. We went on to become friends nonetheless.
(Doug Mills/The New York Times)

It turned out being considered a "wacko bird" by the establishment was considered a compliment, not an insult, outside Washington.

I expressed my opposition to funding Obamacare by filibustering against a budget deal for twenty-one hours from September 23 to 24, 2013. I will never apologize for doing everything I possibly could to stop this disastrous legislation. About eight hours in, when it was Caroline and Catherine's bedtime, I read them *Green Eggs and Ham* from the Senate floor. It was the one thing I had done in the Senate that actually impressed them.

Getting out of Washington—and back to America—kept me grounded. I was honored to be the one who could do what the people wanted, which was to #makeDClisten.

The day after the filibuster, I attended a prayer vigil for Pastor Saeed Abedini, an American citizen who had been unjustly detained by the Iranian regime for the crime of being a Christian. I couldn't understand why President Obama was not making this an issue with the Iranians by insisting Abedini be released before any diplomatic contact with the Islamic Republic. It was an important reminder of how vital—and fragile—our freedoms are. *(Courtesy of the American Center for Law & Justice)*

Saw this, but noticed an error. So I wanted to make one thing clear: I don't smoke cigarettes bit.ly/1nqK08i pic.twitter.com/tPFNqg9vu8

↰ Reply ⇄ Retweet ★ Favorite ▼ Pocket ••• More

One of the consequences of the Obamacare filibuster was that I became, however unwittingly, something of a pop culture figure. On a trip to Los Angeles, we found the street artist SABO had plastered the city with posters of my head on a ripped, tattooed body with the text "Blacklisted & Loving It."

President Obama poked fun at my new reputation at the 2014 White House Correspondents' Dinner, when he did a riff on the legislation I had authored that prevented the Iranians from sending a known terrorist, who had participated in the 1979 hostage-taking in Tehran, to be their United Nations ambassador. It had passed both houses of Congress unanimously. The president showed a slide of the signing ceremony, with him, me, the devil—and hell freezing over. *(Photo Credit: youtube.com/user/ whitehouse)*

I was the only Republican senator to join the congressional delegation to Nelson Mandela's funeral in South Africa. Mandela is a hero of mine because of his passionate fight against racial injustice and his ability to look beyond vengeance and bring healing to his nation after the end of apartheid. But I wasn't blind to the faults of some of his admirers: when Raúl Castro spoke, I walked out.

In May 2014, I traveled to Israel, the Ukraine, Poland, and Estonia—all allies of the United States who expressed considerable nervousness about the state of our relationship. In Kiev we were guided through the Maidan Square by one of the brave teenage student protestors, who had seen her friends shot during the revolution earlier in the year.
(Courtesy of Secure America Now)

Texas has been front and center in the extraordinary energy renaissance that has transformed America in recent years. Rather than being dependent on bad actors like Russia, Iran, and Venezuela, now the future can be fueled by the U.S. and our allies. It will be a wonderful thing if the federal government will just get out of the way.

One of the more pernicious aspects of Obamacare has been the imposition of requirements, such as birth control, on entities for whom this is a violation of faith—from the Little Sisters of the Poor to the Hobby Lobby. I was proud to stand with them.

In the 114th Congress, with the Republicans in the majority, I became chair of the Commerce Subcommittee on Space, Science and Competitiveness. It is an honor to oversee Texas's proud tradition of pioneering and exploration.

Visiting the memorial to the brave men and women who died in the terrorist attack on Fort Hood on November 5, 2009. It was one of my proudest moments on the Armed Services Committee, when I offered legislation authorizing the army to make the victims eligible for the Purple Heart, and it was unanimously approved by the committee. The Purple Hearts were awarded on April 10, 2015.

The 2014 midterm elections came down to two issues: Obamacare and amnesty, and the voters spoke out loud and clear against them. I was proud to campaign with a number of terrific candidates, including this memorable stop with Ben Sasse, who was running for the Senate from Nebraska, along with my good friends Governor Sarah Palin and Senator Mike Lee. Like many others in this cycle, Ben was successful, and we managed to retire Harry Reid as Senate majority leader.
(Courtesy of Ben Sasse for U.S. Senate)

In the spring of 2015, I participated in a round table with Holocaust survivor and Nobel laureate Elie Wiesel to discuss the genocidal threat posed by a nuclear Iran. Truly a great man.
(Courtesy of This World: The Values Network)

With Heidi, Caroline, and Catherine at my side, our family took the plunge, and I announced my 2016 presidential campaign at Liberty University.

Standing on the stage, I couldn't help but think of my parents—one an immigrant fleeing oppression in Cuba, the other a pioneering woman succeeding against all odds—neither of whom could have imagined their son would become a United States senator, let alone aspire to the presidency. The truly amazing thing is that in our country, similar stories happen every day. That is the promise of America.

The election results in Florida on November 7, 2000, had proven both the biggest disappointment and turning point. Veteran CBS broadcaster Dan Rather, a dogged foe of the Bush family, warned his viewers that Texas governor George W. Bush "is by no means out of the race." But it was starting to feel that way. You could see it in the way the pundits were talking on CNN and Fox and the major networks. You could see it in the slightly satisfied gleam in Rather's eyes. In the middle of the evening—while the polls in Florida were still open—CBS called the state for Gore, and the other networks followed suit.

And then something unusual happened. No, not unusual. Extraordinary. Unprecedented, even. On another channel, CNN, the veteran anchor Bernard Shaw appeared on-screen to declare that "CNN is moving our earlier declaration of Florida back to the too-close-to-call column. Twenty-five very big electoral votes." Other networks too were suddenly saying, to obvious embarrassment, that calling Florida for Gore had been oddly premature.

Then Florida swung even further away from the Democrats—and into the Bush column. Just like that, the night—now early Wednesday morning—had taken a breathtaking 180. There was hope again in Austin. No one could sleep. No one wanted to sleep. Then, at 1:18 in the morning Texas time, NBC's Tom Brokaw abruptly cut into a conversation with Katie Couric and presidential historian Doris Kearns Goodwin to declare, "George Bush is the president-elect of the United States. He's won the state of Florida, according to our projections. . . . It's been a night of first giving it to Al Gore, and then taking it away." It made me feel sorry for Gore. Almost.

Even Dan Rather finally admitted Bush had won. With his usual folksy antics, and apologies to Hemingway, Rather called the election for Bush thusly: "The son also rises."

"It's been a long and suspenseful evening," Brokaw said on NBC, as if in summary. Of course, as we all know now, it was just the beginning. A long, ugly mess was unfolding—and at the age of twenty-nine, I was right in the middle of it.

————

When I was hired on the Bush campaign in June 1999, I did not know the governor at all, though of course I knew of him and his famous parents.

Bush first ran for governor in 1994 in what seemed a long-shot bid against the popular Democrat Ann Richards, but he had demonstrated remarkable, almost inhuman, message discipline. He had four priorities, he told people: tort reform, juvenile justice reform, welfare reform, and education reform. He repeated those four over and over again. At one point midway through the 1994 campaign he called some reporters together, "Hey, I have a fifth priority." The reporters grabbed their pads, excited. He said, "My fifth priority . . . is to pass the first four." That was typical Bush humor.

When Bush announced he was running for president, I knew he was the candidate I would support. I believed he had strong conservative instincts. He had been raised in Midland, Texas, which gave him many of his values—a belief in entrepreneurship, free enterprise, and hard work. He had served as governor of Texas for five years, and I admired his record. I thought I might want to help his campaign in some way, but I wasn't really sure how to go about it. At that point I had a great job in private practice in Washington.

Coming out of my clerkship, I had gone to work at Cooper & Carvin, a tiny law firm founded by Chuck Cooper and Mike Carvin, both protégés of Ed Meese in the Reagan Justice Department. Chuck was also a former Rehnquist clerk, a graceful and elegant writer, and one of the top Supreme Court litigators in the country. Mike is simply brilliant, blessed with Scalia-esque intellect and a wicked wit. Theirs was a brand-new firm—just nine months old, and I was just the seventh lawyer to join them. The firm's practice was unique, specializing in cutting-edge constitutional litigation and commercial litigation.

Most clerks go to work for big, established firms, but some of the best advice I had gotten was to work for people that can best teach you how to practice law. I respected Chuck and Mike and wanted to learn lawyering from them. They're principled conservatives, and phenomenal litigators, and both became great friends and mentors to me.

As a baby lawyer, I had just argued my first appeal, a $10 million case representing Ford Motor Company. We won unanimously. I was also helping Chuck represent the National Rifle Association in multiple cases, and assisted in preparing his testimony before the House of Representatives in the impeachment of Bill Clinton. (Chuck testified as a constitutional expert on what constitutes "high crimes and misdemeanors" under the impeachment provisions of the Constitution.)

I also was helping Mike represent a congressman named John Boehner. In 1995, two Democratic activists illegally intercepted a call on Boehner's cell phone and gave the tape to Democratic congressman Jim McDermott, who promptly handed it over to news reporters. Boehner sued McDermott—in what was billed as the first civil suit between sitting congressmen in history—and hired Mike to represent him. Ultimately, the case became a major First Amendment suit, and after many years of litigation, Boehner prevailed.*

In the midst of learning to practice law, I was also dealing with a personal crisis. My older sister Miriam was in real trouble. She and Roxana were my half sisters from my dad's first marriage. Miriam

---

* Although he was my client, I think I only met Boehner once. I was a junior lawyer in the back of the room, and I don't recall that we spoke at all.

in particular had a very difficult time with her parents' divorce. She was beautiful and quite intelligent, and incredibly loving. When I was a toddler, she would play with me for hours on end, even letting me pull her hair, without complaint. But she was perpetually angry—emotionally trapped as a rebellious teenager. Though she was a sweet person with a good heart, she slowly surrendered to her wild side. She would go dancing at what my father derisively referred to as "honky-tonks" and consistently sought out bad characters. She was married to a man who was in and out of prison, and who was physically abusive. Eventually she developed a serious drug and alcohol addiction. I loved Miriam, who was nine years older than I. But I sometimes found it hard to reconcile the bright, fun, charismatic sister I adored with the person who would lie to me without hesitation and who stole money from her teenage brother to feed her various addictions.

As she grew older, her descent continued. Several times, she went to prison for petty offenses, like shoplifting. Yet she never seemed to learn any lessons. Her sickness took a firm hold of her. She would continue to lie, to steal, to commit crimes over and over again.

Coming out of prison, she had hooked up with a man with a serious drug problem, and they moved into a crack house. At that time Miriam had a young son, my nephew Joey, who was in sixth grade. She had been raising him as a single mom, but she was no longer able to do so. My dad flew from Texas to D.C., and he and I together drove to Philadelphia, where they were living. I remember my dad and me leaving our watches, rings, money, and wallets in D.C., because we didn't know if we'd be robbed or shot at the crack house where Miriam was living.

We picked up Miriam and took her to a Denny's to try to talk some sense into her. It was a fruitless effort. She was angry. Angry that her parents divorced when she was a little girl, angry that her father had missed a high school swim meet of hers. I wasn't terribly understanding. I pointed out to her that our aunt, Tía Sonia, had experienced far worse—horrific things in a Cuban jail—and yet she raised our cousin Bibi as a single mom. I told Miriam she needed to stop wallowing in self-pity and do what was needed for her son.

Miriam refused to change her path. And so we had a real challenge in what to do with Joey. My parents were still broke, and I was just starting to work, with substantial student loans. So I took a twenty-thousand-dollar cash advance on my credit card to pay for Joey to go to a military school, Valley Forge Military Academy. It took me several years to pay off the bill, but I think and hope that putting Joey in that school made a real difference in helping instill discipline and order in his life.

A year later, Miriam had improved somewhat and was able to care for Joey again. In 2011, my sister died of a drug overdose, which the medical examiner determined was accidental. It was heartbreaking. I loved my sister, and she spent much of her life trapped by the demons of addiction and anger. During her struggles, my other sister Roxana really stepped up as well, becoming a tremendous source of love and support for Joey. Today, Roxana is a successful doctor in Texas, and she has a real heart for helping and caring for those in need. But I take great solace in the fact that Joey has grown into a strong and responsible young man, working a good job in a Pennsylvania chocolate factory.

Back to 1999. My law practice was thriving, and I was putting down roots, even getting ready to close on my first townhouse. But then, at a fund-raiser for Governor Bush in Washington, I got an unexpected opportunity. A law school classmate introduced me to Bush's policy director (and future White House chief of staff) Josh Bolten. A forty-four-year-old workaholic bachelor whose stubbornly temperate manner masked a razor-sharp intellect, Josh was assembling a team of about a dozen policy advisors for the Bush 2000 campaign.

I had no idea who Josh was, but my law-school buddy whispered, "Ted, talk to this guy. Trust me, you want to talk to this guy."

After a brief conversation, Bolten asked, "Can you come interview with me tomorrow?"

"Sure." I had no idea what I was interviewing for.

The next day, I met with Josh for a couple of hours, and next thing I knew I had a job on the campaign, as one of the domestic policy advisors to George W. Bush. Within two weeks, I had backed out of the townhouse deal, packed up my things, and headed for Austin.*

When I arrived, the campaign was nascent, a small operation on the second floor of 301 Congress, a high-rise building in Austin, less than a mile from the governor's office. Most of the staffers sat in cubicles in a series of big rooms. My portfolio was domestic policy, basically anything that touched on law. It included criminal justice, tort reform, judicial appointments, civil rights, abortion, gay marriage, religious liberty, immigration, gun rights, and campaign finance reform.

Governor Bush intended from the outset to run a campaign that was policy-driven and substantive. There was a lot of pressure on those of us on the policy team to help develop meaningful policy proposals that would underpin the political messaging.

Part of the imperative was that the media caricature of Bush was that he was not terribly bright and was a celebrity candidate coasting

---

*Joining the campaign required an 80 percent salary cut, a sobering decision for a twenty-eight-year-old fresh out of school. My fixed bills, including my student loans, dramatically exceeded my monthly income on the campaign. Fortunately, in my two years at Cooper & Carvin, I had saved nearly half of my salary, so I was able to live off savings during the campaign.

on his parents' reputation. A serious and coherent policy platform would directly counteract that false and unfair narrative.

The policy team was led by three outside senior advisors. Former Federal Reserve governor and Harvard professor Larry Lindsey led economic policy; Stanford provost and Russia expert Condoleezza Rice led foreign policy; and former Indianapolis mayor Steve Goldsmith led domestic policy.

Each had a full-time staff liaison in Austin. Rice's liaison with the campaign was Joel Shin, who became my roommate. Joel and the rest of the team definitely did not fit the media-created stereotype of the Bush team as a bunch of ultraconservative, evangelical white males. Josh Bolten, for example, was Jewish. I was Cuban-American, of course. And Joel was a Korean-American Rhodes scholar from Alabama with three degrees from Harvard and an encyclopedic mind.

For the first couple of months of the campaign, Joel had slept at the office. He kept his suitcase under his desk, showered in the building's gym, and would work around the clock, until he literally slumped over in his chair.

When I arrived, Josh had just given the edict to Joel that he was not allowed to make the office into his residence. So together we rented a two-bedroom apartment across the street from the campaign. In many ways he was the perfect roommate in that he paid half the rent, and yet worked twenty hours a day and would basically walk across the street, shower, maybe sleep an hour or two, and then go back to the office.

As a leader, Josh taught us an important lesson in management: He didn't husband access to his principal. He let each of us on the policy team engage directly with Governor Bush, and answer his questions.

Many people, especially those who were veterans of Capitol Hill, would have behaved differently. One of the keys to success on the Hill is to control access to those who hold power, and so a lot of staffers try to shut their principal off from having interaction with anyone else. Josh understood that if he had a dozen experts spending all of their time delving into different subject matters, each of them could be far more informed on a particular subject than he possibly could (although he was awfully well informed himself). Granting them full access was a better service to his boss.

We would usually have an hour or two of policy time each week scheduled with the governor. We'd go to his living room at the Governor's Mansion, rather than his office, because he couldn't campaign from a government office.

Bolten also encouraged us to take personal ownership of our work. He instructed us to put our names on the bottom of our memos—which had to be succinct—as well as our cell phone numbers so that the governor could call us with any questions. Unfortunately, he tended to exercise this option at what I considered to be an inconvenient hour. Indeed, one of the areas where he and I were at odds from the beginning was that George W. Bush is a morning person.

Usually around 6 a.m., I would be dead to the world, having gotten home from the office at two or three in the morning. When the phone rang that early, I knew who it was—nobody else who knows me would call me that early—so I'd sit up in bed, reach back, and slap myself across the face as hard as I could to wake myself up.

Bush was not someone who engaged in a lot of chitchat or stopped to ask how your morning was going. His opening question would be something like, "So, is Congress going to pass this thing?" Half asleep, I would be confused enough that I'd almost wonder, "There's a Congress?" Then I would have to remember what I had written a memo on so I knew what the "thing" was.

Over the course of the next year and a half, I came to know Governor Bush fairly well. Perhaps because I came off as super-serious, his nickname for me was "Theodore." I don't know if he assumed that was my full name, which it is not, but it didn't really matter for his purposes. I came off lucky next to Karl Rove, whose nickname, famously, was "Turd Blossom."

Bush has remarkable charisma, particularly one-on-one or with a small group of people, which is invaluable to a politician. He's one of the most natural people-persons I've ever seen. The key component to that charm is his self-deprecation. We once were having a discussion over the death tax, also known as the inheritance tax, which is obviously of significant concern to families with a lot of money. Bush bluntly said, "The best cure for family money . . . is a third generation." It was hard not to laugh at that, and also to be impressed, since Bush was a third-generation grandson of the wealthy Prescott Bush and, as was widely known, his track record in business had not been an unmitigated success.

Throughout the course of the campaign I was also impressed with Bush's inquisitive approach to the issues at hand. He never accepted our policy papers at face value. He would ask hard questions of us, and those questions seemed to me to spring from conservative principles. He believed strongly in freedom as the natural state of human beings, as God's gift to humanity. Throughout his presidency, his deeply held beliefs made him an instinctive soulmate to Wiesel, Natan Sharansky, and other dissidents across the world.

Bush's conservatism was a different sort from that of, say, Ronald Reagan. Reagan had spent a lot of years thinking about first principles, reading Milton Friedman and free-market economists. He was someone who addressed public policy issues from his core positions and reasoned from there.

Bush's was a sort of gut conservatism shared by a lot of small business owners in West Texas. Although he was bright and asked incisive questions, he didn't tend to reason from abstract principles. Nevertheless, I got the feeling that at any campaign meeting, perhaps with the exception of a few junior staffers like me, Bush was always the most conservative person in the room.

That, as it turned out, was a problem. Bush often found himself being urged by pointy-headed policy advisors to tack to the center, or even the left, on a variety of issues on the grounds that it would make him seem more "reasonable" and less of a Texas cowboy whom the advisors feared might look scary to the electorate. Throughout the campaign and in the early years of his administration, Bush tended to resist that pressure.

One of the first policy sessions I sat in concerned Social Security reform. Virtually every staff member in the room was advising Bush to stay away from the subject, following the conventional wisdom that Social Security is the proverbial "third rail" of politics: If you touch it, you'll get electrocuted. This was, I soon learned, the prevailing sentiment in Washington, D.C.—avoid anything controversial or courageous. One staff member in the room said, "This issue polls terribly." In his opinion, that should have been that.

There were few things you could do to tick Bush off more than make an argument based solely on opinion polls. In that meeting, Governor Bush pounded the table. "Damn it, this is the right thing to do," he said, and asked us to come up with a specific proposal. There were a number of issues on which I disagreed with Governor Bush, both as a candidate and as a president, but one thing that I will very much credit him with is the courage of his convictions, that he was willing to stand by what he believed, even in the face of daunting odds.

I stayed quiet during that discussion, because I'd just started on the campaign. But it made a real impression on me. He told the team, "It's your job to give me the best advice on what the right policy is. It's my job to figure out how to sell it." I agreed with that proposition: Substantive policy should not be derived by polling.

Over and over again in the campaign, anytime a policy person invoked polling, it provoked an immediate and negative visceral reaction from Governor Bush.

Now, that's not to say polling had no legitimate role. Karl Rove and Karen Hughes were both looking to polling in terms of how to describe an issue. Obviously Bush's goal was to get elected, and he couldn't completely ignore their data. But in terms of arriving at the substantive policy position, polling was never the driver.

Bush's views on immigration during the campaign were quite different from what ultimately materialized as the administration's comprehensive immigration plan, which caused such controversy in 2007. The focus on immigration in the campaign was twofold. In the campaign, the focus was on securing the border and improving and streamlining legal immigration. There was not an amnesty component to the 2000 proposal.

One of the many things Bush understood was that with immigration, tone matters. In Midland, Bush grew up surrounded by the Hispanic community, and he understood their concerns in a real way that many Republican politicians simply don't.

This was brought home to me during the 2008 campaign when I joined a number of other Hispanic Republicans from Texas in attending a Republican debate in Miami that was sponsored by Univision. At times it felt to us like the candidates were basically saying, "*You people* need to vote for us."

I and the other conservative Hispanics around me were looking at each other going, "Who, exactly, are 'you people'?"

———

The Bush 2000 campaign was exceptionally well run, but it's most frequent failing was a continuing sense of entitlement. Over and over again, when we were ahead in the polls, we'd relax and stop drawing meaningful distinctions. Our ads would focus instead on fuzzy pictures of kids, saying essentially "we like children" (as if somehow the other candidates were anti-kid).

This is the wrong way to run. As the old political saying goes, there are two ways to run: scared and unopposed. We were neither.

In early 2000 Bush won the first presidential contest, the Iowa caucuses, and so the campaign promptly relaxed. It was as if we expected some sort of coronation. In the following contest in New Hampshire, we didn't run a hard-hitting campaign—on the day of the primary, John McCain held a rally with thousands of people, and Bush went sledding with his family—and we paid the price. We had played the equivalent of a "prevent defense" in football, and as a result we got our teeth kicked in, losing to McCain in New Hampshire by 19 points.

That was the best thing that happened to the Bush campaign. Finally, the senior campaign leadership decided to draw serious distinctions between Governor Bush and his opponent. As it happened, there were two issues on which Senator McCain's record was substantially to the left of George W. Bush's: campaign finance reform and tort reform. Both were in my portfolio.

Until the South Carolina campaign—which followed the New Hampshire primary—I had largely been confined to Austin with most other policy staffers. In South Carolina, however, I spent the entire week before the primary on the road with Governor Bush. We went city to city in a decked-out Greyhound bus and for the first time I saw Bush up close for a prolonged period.

I quickly understood Bush's advantage over McCain. Bush was a Texan, and he connected easily with voters in a southern state like South Carolina. Like Texas, South Carolina is home to a lot of military families, evangelicals, gun owners, and cultural conservatives. As we traveled the state, Governor Bush's message resonated with those voters, and the two policy issues that he rolled out there made his differences with John McCain crystal clear. Bush ended up winning that primary by a substantial margin. And he was back on the road to the nomination and the tension-filled, close-as-could-be general election that followed it.

————

Election day was somewhat surreal. The policy team, which had worked twenty hours a day for a year and a half, was suddenly irrelevant. Nobody cares about policy on election day.

And so we went bowling. With Bo Derek.

As it so happened, my boss, Josh Bolten, was a big devotee of bowling. And he was dating actress Bo Derek. They had met at the 2000 Republican National Convention and had curiously sparked up a romance.

So when Josh took the entire policy team bowling on election day, Bo joined us. Let me just say, Bo Derek is spectacularly beautiful. In person, she is even more stunning than on the silver screen. At the bowling alley, she bowled barefoot, with two hands, in a white pantsuit. Every man on the policy team was mesmerized.

Indeed, later that night, as we were walking home from the never-ending election results, I paid the price for that afternoon. My girlfriend and future wife, Heidi, asked me the seemingly innocuous question, "So, do you think Bo Derek is pretty?" Sober, I can answer that question just fine. But that night, at four in the morning, I wasn't sober. And my response had far too much enthusiasm. To this day, Heidi gives me (justifiable) grief over my effusive response to her query.

Two days after the general election, with the Florida results still in doubt, Josh called me to his office. Even in this crisis, he spoke in his customary calm, business-as-usual cadence.

"Ted, I'm going to need you to go down to Florida for me," he said. "Right now."

There was just enough in his tone to suggest the situation was urgent. Without packing anything I rushed straight to the airport, jumped on a plane, and arrived in Tallahassee that very afternoon.

Almost everyone in Austin thought the recount would last a couple of days. This would be a quick, perfunctory legal proceeding. The votes had been counted, we figured. Bush had won.

Indeed, I was astonished that Al Gore had even decided to challenge the election in the courts. In the history of the country there had never been such a legal contest of a presidential election. Even Richard Nixon, who lost to John F. Kennedy in 1960 amid widespread allegations of fraud in Chicago and Texas, resisted the urge to contest the results and divide the country indefinitely. I thought it was a rather petulant display by Vice President Gore.

In our headquarters in Tallahassee there was only a skeleton staff as yet. I immediately connected with Ben Ginsberg, a longtime friend and the lead outside counsel to the Bush campaign. The forty-eight-year-old Ginsberg, with a thin, reddish beard and a bald head, was a seasoned election lawyer. I knew we were in good hands.

Other lawyers in his position, with the eyes of the world upon him, would have soaked up the media spotlight. But what Ginsberg did was instructive. He knew what most of us did not—that this was not an open-and-shut case, and that it would be headed to court. He recognized that his specialty was not litigation, so he didn't say, "I'm the leader of this battle." He instead said, with great humility and restraint, "We're going to need a lot of lawyers to help on this one." He was more than willing to surrender his ego to the cause, and to bring in even bigger dogs to take the lead in the case. And the biggest big dog there was—at least in terms of Republican politics—was about to arrive.

———

When he ran for president, George W. Bush made a point of differentiating himself from his father, former president George H. W. Bush. At one point he famously looked around a room of advisors, and, referring to prominent members of his father's administration, noted "Look who's *not* here." The candidate wanted everyone to know he wouldn't turn to his dad's friends to bail him out of a jam or to water down his conservative policy agenda.

So it was a sign of how serious the Florida recount situation was that Bush very quickly tacked in the opposite direction. Wisely so. He called on the most prominent, effective member of the first Bush administration—his father's longtime friend James A. Baker III.

Baker had served as President George H. W. Bush's secretary of state and campaign manager. Before that, Ronald Reagan had made him chief of staff and later secretary of the Treasury. A highly effective, charming, and shrewd negotiator, Baker had played a leading role on five presidential campaigns.

Tapping Baker was a brilliant pragmatic move by George W. Bush. In a battle that was both the Super Bowl of litigation *and* the World Series of politics, Baker had the mind of a skilled lawyer, the credibility of a statesman, an easy rapport with reporters, and a well-earned reputation for ruthlessness masked by Texas charm. All of that made him an ideal leader for our legal team.

———

When it became clear we were in for a long fight, Ben Ginsberg and I sat down together with a notepad and started to discuss whom we should ask to come down and help us.

The first person I suggested was my former boss Mike Carvin. There's no one better if you are in a fast-paced, unpredictable litigation. With blazing speed, he can think through all of the difficult tactical decisions likely to occur.

When I called Mike, he was at a wedding in Seattle. I said, "Mike, we need you to come down here." He didn't hesitate. "I'll be right there," he said. Mike got on a plane and flew to Washington, D.C., where his wife met him at the airport with a suitcase full of clothes. Hours later, he was in Tallahassee.

Another friend I called was the head of Supreme Court litigation at the venerable firm of Hogan & Hartson. It was a fellow named John G. Roberts. While I had been clerking for Chief Justice Rehnquist, the Chief mused to me once that he thought he could get a majority of the Court to say that John was the best Supreme Court litigator in the country. Like Mike, John understood what was at stake in Florida. He jumped on a plane and immediately came down.

Roberts was not only a brilliant Supreme Court lawyer, but startlingly low-key and self-effacing. Although he was one of the leading Supreme Court litigators in the nation, he had befriended me several years earlier when I was a baby lawyer. He didn't have to do so, and most people in his position would not. For that, I have always been grateful.

One day about midway through the recount, he was leaving the headquarters of the Florida Republican Party—the George Herbert Walker Bush Building, as it so happened—and was carrying a suitcase. I said, "John, where are you going?"

"I've got to get back to D.C.," he said.

"John, we're in the middle of a battle for the country," I said. "How can you be leaving?"

He somewhat sheepishly replied, "Well, I've got a U.S. Supreme Court argument tomorrow morning."

I had to laugh and say, "Well, that's a pretty good reason for you to go."

Roberts flew home, argued a complicated intellectual property case that he went on to win, 9–0, and then immediately returned to work, helping us win the recount battle.

————

Some of the depictions of the recount have described the Bush team as carrying out an elaborate, detailed, meticulous legal strategy. That's nonsense. It was one step shy of utter chaos. I should know, because I helped managed the process.

Over the thirty-six days, there were seven different legal proceedings that could have cost George W. Bush the presidency of the United States. There was no blueprint or guidebook for that seven-ring circus. It was a madhouse. Indeed, in the first six days I was in Florida, I slept a total of seven hours.

To give just one example of the fly-by-the-seat-of-our-pants nature of the endeavor, I can remember tearing pages from legal briefs just hours before they were due. The lawyers who wrote them would protest, "We've gotta file this in two hours! What the heck are you doing?"

"I don't care," I would reply. "You can't say this! We said the opposite thing in two other courts!"

In retrospect, it was pure serendipity that I found myself in the middle of this extraordinary legal proceeding. As it happened, I was the only practicing litigator who'd been on the full-time campaign staff. Moreover, the legal issues were constitutional, and I'd practiced constitutional litigation. And, my background was at the U.S. Supreme Court, and that's where this legal fight was heading.

That's not to say that others didn't see me as too young for the job. Once, over Thanksgiving, I was walking over to the federal district court in Tallahassee to file a pleading. One of the more senior Bush attorneys (who later became a cabinet member) grabbed me by the shirt, pulled me up in his face, and growled, "Ted, don't f*** this up."

I calmly replied, "It entails walking across the street and handing these papers to the clerk. . . . I feel confident I can accomplish that task."

Everyone was on edge and had good reason to be skeptical. This was the finest litigation team ever assembled, a legal dream team, and it included past and future cabinet members, deputy cabinet members, and White House chiefs of staff. I felt like Opie on *The Andy Griffith Show*.

Regardless of age, to a person every one of the lawyers on our team was horrified at what was happening. We sincerely believed that our candidate had won the election and that the other candidate was trying to steal the presidency.

I believe to this day, however, that there was a big difference between our approach and theirs. The fact is that every time the ballots were counted—*four separate times*—Bush won. The Democrats' approach was simple: They were going to keep counting, and counting, and counting until Gore won. But he never did, and at some point they had to stop counting.

———

My former law professor Alan Dershowitz, who supported the Gore campaign in that proceeding, has said, "The Bush team emphatically outlawyered the Gore team."

Part of the reason is very simple. We had senior, experienced lawyers handling every component of it. Then we had senior press surrogates carrying the public messaging, most significant Jim Baker.

On the Democratic side, almost every one of those jobs was done by David Boies. An extraordinarily talented attorney, Boies is one of the most versatile litigators I've ever seen. He is great with the facts. He is great with the law. But there are limits to human endurance.

Boies, it seems, failed to recognize the simple reality that he had only twenty-four hours in his day. He led the federal litigation team for Al Gore, the state litigation team for Al Gore, and the U.S. Supreme Court argument for Al Gore. On top of that, he became Gore's chief press spokesperson.

The last job he got by default. The first two press surrogates Gore relied upon were Warren Christopher and Bill Daley. Christopher was an aging former secretary of state who did not come across well on camera. And, through no fault of his own, Daley's image was even worse. It simply was not ideal to have the scion of one of Chicago's great political families insisting, "We're not stealing an election."

In any other litigation, the performance of the Democratic lawyers at trial would likely have been deemed malpractice. I don't say that lightly.

One example suffices. At the contest trial Gore's team presented just two witnesses. One of them was a statistician from Yale. The Gore legal team's theory centered on partially punched "chads," which are perforated squares on the ballot that voters punched with a stylus in order to select their candidate. The statistician was supposed to show that votes for Gore went uncounted, because over time too many chads would stack up under the first column of the punch card machines, making it harder for Gore voters to push through their vote in that column.

The theory never made much sense (presumably the "problem" would have affected Bush voters, too). Regardless, to support their theory, the Yale statistician testified that in a previous election, there had been significantly more votes cast in the U.S. Senate race than in the governor's race. The only explanation for this difference, he testified, was that the candidates for governor had been in the first column of the ballot, while the candidates for Senate were in the second column.

Thus, the Gore team argued, the chads from the first column had built up in the voting machines, and voters who wanted to vote for the governor could not do so. The same thing, their theory went, happened with Gore in the presidential race in 2000. Our campaign's trial attorney, Phil Beck delivered a devastating cross-examination of the statistician that proved that Gore's star witness hadn't even examined the ballot. It turned out that *both* the governor's race and the Senate race were in the first column, which forced the statistician to admit on the stand that his entire testimony was flawed. It was the most devastating cross-examination I've ever seen.

But our strategy was hardly rocket science. Rather, we did the basic legal homework that lawyers should do to prepare for court. We had read the statistician's witness report, pulled the actual ballot, and discovered that he didn't know what he was talking about.

The Democrats should have known that as well. If there's one thing that the Democratic Party has in abundance, it's capable trial lawyers. For the life of me, I cannot understand why they did not recruit a serious, substantial legal team for every aspect of the recount.

———

How does one count a partially punched chad? That question bedeviled the Florida courts.

Unfortunately, some voters did not properly punch the chads for their preferred candidates. For example, some voters punched the chads by Al Gore's name *and* by George Bush's name. Some voters partially punched Gore's chad and completely punched Bush's chad. Some voters partially punched Gore's chad and did not completely punch any chad. Some voters partially punched Gore's chad and

Bush's chad, but punched one more forcefully than the other. And so on.

Before long, people were calling a chad with three of its four corners punched a "hanging chad." A chad with two of its four corners punched was a "swinging chad." A chad with all four corners attached but with a dimple in the middle of it was a "pregnant chad." It all became a bit silly.

As the litigation progressed, a key legal question arose: In a statewide recount, is it constitutional if each one of Florida's sixty-seven counties uses a different standard for what counts as a vote? In other words, in a statewide recount, is it constitutional if Palm Beach County counts hanging and swinging chads, while Broward County counts only swinging chads, and Miami-Dade County counts hanging and swinging and pregnant chads?

Gore said yes. We said no. In our opinion, a presidential campaign could not be decided by sixty-seven counties saying, arbitrarily, "We'll count this, you'll count that, and it doesn't matter what the next county counts." That standardless chaos, we argued, was inconsistent with the Fourteenth Amendment, which guarantees "equal protection of the laws" and thereby prohibits the government from arbitrarily treating different groups of people differently.

The Florida state courts disagreed with us. So ultimately, we asked the U.S. Supreme Court to answer the question.

When we first decided to take our case to the U.S. Supreme Court, there was a division of opinion among the lawyers as to whether the Court would take the case. Personally, I believed the justices would see political risk in getting in the middle of such a contentious and political dispute. But having clerked on the Court, I also believed they would feel a responsibility not to duck such an important case.

As we prepared for our first trip to the Supreme Court, I worried that our lead argument was too aggressive. I suspected the justices would be cautious about making a sweeping decision in such a case. Often in litigation, swinging for the fences is a mistake; one of the surest paths to victory is finding a narrow ground that benefits your client but doesn't require the Court to go too far out on a limb.

So I suggested a fallback argument: We should tell the Court that even if they didn't completely agree with us, they should nonetheless vacate the decision below, clarify the federal law in question, and then send the case back down to the Florida Supreme Court to reconsider. I worked with Tim Flanigan and Noel Francisco—two dear friends, and incredibly talented lawyers—to draft the portion of our brief asking the Court to do so.

And, in *Bush v. Palm Beach County*, that's exactly what the U.S. Supreme Court ended up doing. Unanimously.

The highest court in the land remanded the case back to the Florida Supreme Court . . . which then proceeded, astonishingly enough, to ignore the U.S. Supreme Court. Instead, the Florida Supreme Court (consisting of seven judges, six of them appointed by Democrats) simply reinstated its prior order of a statewide recount that would allow each of Florida's sixty-seven counties to use a different standard for deciding which hanging, swinging, or pregnant chads count as "votes."

Because we believed the Equal Protection Clause of the Constitution was violated by a recount that did not treat the votes of one county equally with the votes of another county, we went back to the U.S. Supreme Court.

In oral arguments there, our lawyer, Ted Olson, emphasized that the Florida Supreme Court on remand, after being unanimously vacated by the Supreme Court, did not so much as cite the U.S. Supreme Court decision. They just ignored it. If there is anything that the Court's swing justices—Justice O'Connor and Justice Kennedy—believed in, it was the authority of the U.S. Supreme

Court. Both of them were outraged by the Florida court's defiance. Every time Ted reminded them of that defiance, Justice O'Connor or Kennedy would say something to effect of, "What? They didn't cite it?? What?!"

I don't think the Supreme Court's outrage affected its 7–2 decision on the merits, agreeing with our argument that the statewide recount ordered by the Florida Supreme Court was unconstitutional because it allowed different counties to count votes using completely different standards. But the justices' outrage did affect what came next.

Two of the seven justices who agreed with us were willing to give the Florida Supreme Court yet another chance to get it right. But five of the seven justices who agreed with us were so outraged by the Florida Supreme Court that they said, in effect, "Enough is enough. This is a presidential election. It has dragged on for thirty-six days. The ballots have been counted four times. Bush has won every time. He is the winner. The end."

―――――――

On December 12, at about 10 p.m., I got a call on my cell phone from the clerk's office of the U.S. Supreme Court. The caller, an old friend from my days at the Court, told me, "The decision is coming down. We're going to fax it to you now."

I pulled the papers off the fax machine and carried them into Jim Baker's office. As he stood there wearing a dark green Michigan State sweat suit, I handed him the opinion.

He handed it back. "Well?" said Baker. "What does it say? What does it mean?"

It was a complicated, twenty-five-page opinion, and despite my paraphrasing of it above, it did not include a line that said, "George W. Bush is president. The end." Indeed, at that very moment, reporters were standing on the steps of the Court, utterly befuddled trying to figure out the ruling.

I read over it for several minutes, as quickly as I could, with the former secretary of state looking silently over my shoulder.

After a few minutes, I looked up at Baker and said, "It means it's over. We've won."

Secretary Baker looked at me, nodded, and placed a call to Crawford, Texas. "Well, Mr. President," he drawled, pausing slightly after he used that title for the first time, "how does it feel?"

Chills ran down my spine.

Weeks later, Heidi observed, "It's a good thing you were right!" It then occurred to me, "Holy cow, what if, say, in footnote eleven, it had given the Florida Supreme Court and Al Gore one more crack at the apple?"

––––––––

Coming out of the campaign, I hoped to get a senior job in the White House. I thought I had done a good job on the campaign and that the president-elect had appreciated my work. But I didn't get it, and for a simple reason: In the heat of the campaign, I had forgotten some of my own life lessons learned during my seventh-grade makeover.

Instead I was far too cocky for my own good, and that sometimes caused me to overstep the bounds of my appointed role. I foolishly thought it was my job to provide my best judgment on the right policies for our candidate. I didn't understand that lots of others on the campaign thought my job was simply to be a conduit for their own expertise. They really didn't give a flip what some twenty-something kid thought might or might not be the right policy outcome. As a consequence, I burned a fair number of bridges on the Bush campaign.

So, instead of going into the White House in the Bush administration, I served as one of four members of the transition team for the Department of Justice. I then became an associate deputy attorney general, with responsibility for legal policy and legal counsel.

I was at the Department of Justice only briefly, in part because I always felt like an outsider there. Attorney General John Ashcroft had brought over his own team from his Senate office (he had served as U.S. senator from Missouri) and I never seemed to be able to earn their trust. At one point, folks from the White House related to me that the Ashcroft team complained that I was too "loyal" to Bush. It had not occurred to me until then that supporting the president was deemed inconsistent with the workings of the Bush Justice Department.

I enjoyed working with the career prosecutors at DOJ—men and women deeply committed to the rule of law—and grew to very much respect the integrity of the department. By far my most interesting experience at DOJ was my brief but telling tutorial on U.S.-European relations. In 2001, I helped lead the U.S. delegation to Rome for the Council of Europe's negotiations on a treaty relating to cybercrime.

My sojourn in Rome played out almost exactly according to preexisting stereotypes about the nations of Europe and their relationship with America. For example, almost every time the United States espoused a position, the delegate from France would oppose it reflexively.

The French were masters at posturing. The negotiations were mostly conducted in English, but the French delegate, who was fluent in English, would speak only French. Those of us who were not fluent (two years of junior high French didn't count!) quickly had to reach for our earpieces for the translation.

Conversely, the British and the Canadians could both be counted on to support us steadfastly in negotiations. This led me very quickly to realize that the most important country affecting the dynamics was Germany.

The German delegate was a large man with a thick beard—almost out of central casting—and he proved to be a fair arbiter. He didn't come to Rome with an obvious ideological axe to grind or a particular country bias. So what I endeavored to do the rest of that week was go and press the case to the German delegation. If they agreed with us on the merits, then practically everyone else would as well. That in fact is exactly what happened.

Ultimately, we finalized the terms of an important treaty on cybercrime, something that has become all the more important in our modern world. Whether it is the North Koreans hacking Sony to try to stop the release of a movie making fun of Kim Jong Un or ISIS hacking U.S. Central Command's Twitter feed to spread Islamist propaganda, cyberattacks are becoming a greater and greater threat to our nation. And we will need serious tools, and cooperation with our allies, to protect ourselves going forward.

———

After I had spent several months at the Department of Justice, Tim Muris, the newly named chairman of the Federal Trade Commission, reached out to me. I'd gotten to know Tim on the Bush campaign. One of the campaign's many outside advisors, he was a seasoned lawyer and economist.

Tim had deep experience relevant to the FTC, whose statutory mandate is to enforce federal antitrust and consumer protection laws. He had served as the head of the FTC's Bureau of Competition under President Reagan and had also served as head of the Bureau of Consumer Protection. He was looking for someone to lead his policy planning team, and asked if I wanted the job.

The move from the Justice Department to the FTC was not typical. In the ordinary hierarchy of Washington, Justice is considered a more prestigious agency than the FTC. But I was happy to go to the FTC, for a number of reasons.

First among them was that I liked and deeply respected Tim Muris. He said he wanted to use the FTC to advance strong free-market principles, unlike some of his predecessors in Democratic administrations, who used the FTC to try to expand government control of the economy.

Second, Tim envisioned my policy planning role to be innovative and aggressive. I wasn't able to be either of those things in the Justice Department. So, toward the end of May 2001, I left for the FTC.

But I didn't go to work right away, because I had a more pressing engagement. I was getting married.

————

Historians may differ, but in my opinion the most important moment on the Bush campaign was when they hired a twenty-seven-year-old Harvard Business School student named Heidi Nelson.

During Heidi's second year at Harvard, one of Heidi's mentors suggested she head down to Austin to work on the Bush policy team. She had at the time a world of options—she had worked on Wall Street for a number of years and had done well at Harvard—but thankfully she prayed on it, and chose to come to Texas for the length of the campaign.

Heidi is a brilliant, meticulous, sunny blonde from California, and I was smitten with her almost immediately. From our first date—a nearly four-hour dinner at a place in Austin called the Bitter End—we found threads in our individual lives that tied us together. For one, there was the profound love and respect we both had for our fathers, each of whom had survived a challenging youth.

Heidi's father, Peter, had grown up in an alcoholic home in Los Angeles. His father was a urologist who abandoned the family for one of his nurses. As a result, Heidi's father was raised by a single mom and he had to help raise his younger sister. He learned responsibility at a very young age, and with great love, he taught that personal responsibility and self-discipline to his own children.

When Heidi was a little girl, she lived for several months in Kenya and Nigeria, where her parents were Christian missionaries. They're Seventh-Day Adventists, a Protestant denomination that strictly observes the Sabbath from Friday sundown until Saturday sundown. As a child, she would play for hours in the African wilderness, with kids who spoke only Swahili; somehow the children could all communicate regardless of language barriers.

Heidi's maternal grandfather had spent more than thirty years as a missionary in Africa; he was a physician, and helped build a small village. And Heidi's brother, to this day, is a missionary and orthopedic surgeon, providing medical care for children in Haiti and the Dominican Republic.

Being raised in a missionary family imparted a strong sense of service to Heidi, and a burning desire to make a difference in the lives of those who are less fortunate.

Most of her childhood was in California, where, at the behest of her father, Heidi started her first company with her brother when she was six years old. That's not a typo. She and her brother baked bread every day after school from 4 to 8 p.m. They repeated this task every day except Saturday, which was the Sabbath. Then on Sunday they sold the bread all day long at a local apple orchard.

Quickly Heidi's small company—which she called "Heidi's Bakery"—became a real business, selling as many as two hundred loaves a week. By the time she was a teenager, she had saved up somewhere in the vicinity of fifteen thousand dollars. This, as it turned out, was money Heidi needed by the time she applied to college.

Heidi's father is loving, but very strong-willed. He's a dentist and a driven athlete, and for many years he was a mountain climber; in 1990 he climbed Mount Everest, nearly losing his life just a few hours short of the summit. He had gone to Adventist schools, and he insisted that Heidi do so as well.

Heidi disagreed. She had her heart set on Claremont McKenna, a small liberal arts college in Southern California. And when Heidi decides on something, she cannot be dissuaded. At first her father refused to pay tuition. She battled with him, and paid her own freshman year tuition with the money she had saved selling bread as a child.

The story immediately struck a powerful chord with me. My mother had likewise stood up to her father to go to Rice. Heidi's situation was very different from my mom's—her father shares no characteristics with my alcoholic grandfather—but I immediately connected with and admired Heidi's strength of character and de-termination.

We began dating three days after she arrived on the campaign, and very quickly became best friends.

After the campaign, Heidi came down to Florida for the recount. She wasn't a lawyer but she wanted to help the effort any way she could. One day she went into the office of Robert Zoellick, who was serving as Jim Baker's de facto chief of staff. Sitting at his desk with his glasses perched on the tip of his nose, Bob peered up, and Heidi said, "Bob, I just wanted to see, is there anything I can do to help?"

He said, "Yes. Grapefruit juice. I want grapefruit juice." And with that he went back to work.

Heidi came into the office where I was working hopping mad.

"Damn it," she said. "I've got a Harvard MBA. I've worked on Wall Street as an investment banker. And his request for me is grapefruit juice!?"

After a moment, she asked me, "What do I do?"

I sympathized with her completely. Then I said, "Sweetheart, here are the car keys. Go get him grapefruit juice, right now."

Heidi sighed heavily, took the car keys, and drove to the grocery store. She bought a little cooler, filled it with ice, and put a container of grapefruit juice in the cooler. When she came back, she set it on Bob's desk with a little Post-it note attached: "Per your request, grapefruit juice."

The next day she went into his office again and asked hopefully, "Bob, anything I can do to help?"

He looked up at her and said, "Raisin bread. I'd like some raisin bread."

This time she didn't need to vent. She just took the car keys, bought him raisin bread, and didn't think about it at all.

After Bush became president-elect, he named Bob Zoellick to the cabinet as the U.S. trade representative. One of the very first phone calls Bob made was to Heidi to ask her to be his special assistant.

"You know, for a week I'd been asking people for grapefruit juice," he observed. "No one would get it for me, and you actually got it done."

Heidi told that story in a commencement address to illustrate to the graduates that no job is beneath you. In fact, when she started as his special assistant, Bob said, "I don't like the trash can in my office. It's square. I want a round one." In the same breath he continued, "and also, I need you to develop a small business policy for international trade." It wasn't long before she'd solved both problems.

The whole time I was in Florida for the recount, people were harassing me about asking Heidi to marry me—in fact, it was all I wanted to do. I had planned to ask her in Austin a few days after the election. But I had to leave for Florida so quickly I didn't even have time to pick up the ring at my apartment. Ted Olson's wonderful wife, Barbara (an amazing woman who was tragically killed aboard the plane that hit the Pentagon on 9/11), needled me for an entire flight about not missing my chance with such a great girl. I couldn't have agreed with her more, and the greatest relief for me once the election was finally over was to get down on one knee and finally ask Heidi to marry me.

She didn't respond the way I had envisioned. She burst into laughter at what I had thought was gallantry. But she said yes!

In May 2001, Heidi and I got married in Santa Barbara, California, just south of where she had grown up. The day before the ceremony, we took our wedding party to a picnic at Rancho del Cielo, the Reagan ranch (it's now a museum). I think for most of the wedding party, that was their favorite memory of the weekend, but I would be in deep trouble if I were to say it was mine.

I will say it was a moving, even spiritual experience. The ranch featured a modest home of about 1,500 square feet. I remember standing behind the chair at the dining room table where President Reagan would do much of his work looking out the window at Lake Lucky. I would not sit in the chair, would not dream of doing such a thing. But I stood probably twenty or thirty minutes behind that chair, looking out that window and soaking up the ambiance of a man I've admired my whole life for having the courage to stand by his deep principles and the ability to lay out a vision that transformed this country and the world.

After the wedding, Heidi and I took a two-week honeymoon in St. Thomas. It was magical. We walked the beach every sunset, bodysurfed, and took a boat ride to nearby islands. After a year and a half on the campaign, the recount, the transition, the beginning of the administration, we were also both exhausted. The first day, we closed the drapes and slept sixteen hours. For the next two weeks, we slept between twelve and fifteen hours every night. In between, we relaxed with each other, with no cell phones and no email. Heavenly. Then we both went back to work.

———

I think you could write a book on how to run a federal agency based on how Tim Muris led the Federal Trade Commission. But don't worry, this is not that book.

A mistake that many Republicans make in government is to view the agency they're heading as the enemy. They view their mandate as stopping bad things at the agency. But Tim Muris understood that bureaucratic inertia is a powerful force. It is like fire; if you fight it directly, it has the potential to consume you. Tim taught me it is far more effective to shape and direct the focus rather than directly attack the career professionals.

Most of those career professionals are good, decent, honorable people. Naturally, they are less than pleased when some political leader comes in and says, "Everything you've done with your life has been harmful. Stop!"

Tim developed an aggressive positive agenda for his agency. The career professionals could focus on beneficial things, although they were often different things than one would see in a Democratic administration that constantly sought to expand government control over the economy.

Tim and I were both strongly influenced by Robert Bork's 1978 classic, *The Antitrust Paradox*. Bork decried the direction of antitrust law for empowering the government to pick winners and losers between competing businesses. It is not, Bork argued, the job of antitrust law to protect *competitors*. It is the job of antitrust law to protect *consumers*.

Bork believed that antitrust law should be concerned not when competitors battle it out in the marketplace, but when they put down their swords and embrace arm in arm in a cartel, which allows them to cooperate with each other against consumers. That may be good for the competing companies, but it's harmful to consumers. In 1978, what Bork was saying was radical and revolutionary.

During my tenure at the FTC, we would constantly ask ourselves: What can we do to expand competition? To help consumers? In trying to answer these questions I learned a great deal about the anticompetitive tendencies of large corporations and industries, and about the proclivity of government to favor big business.

For example, at the FTC I created a task force on e-commerce, with an eye toward encouraging competition on the Internet. We held three days of public hearings, considering ten different industries. In each industry, existing brick-and-mortar providers had come to government seeking barriers to entry, to prevent new competition over the Internet.

Take contact lenses. All across the country, opticians were trying to create barriers to entry for "1-800-CONTACTS," ostensibly because of health risks to consumers. But the principal motivation was that the eye doctors had a ready-made group of customers in their patients, and they didn't want them going to the Internet for cheaper lenses.

Another instance where we intervened was in litigation in Oklahoma between funeral directors and online casket sellers. The funeral directors did not like e-competition, because caskets have a huge markup. So they argued it was in the consumers' interest for only morticians licensed in each state to be able to sell a casket.

Consider also wine. It used to be if you wanted to have a winery, you had to have a distribution system. But now, with the Internet and shippers like FedEx, someone can grow grapes, make wine, and sell it without a single truck, warehouse, or wholesaler. In fact, Texas has grown enormously in this area, developing dozens of new wineries, which has created thousands of new jobs.

The problem in the early 2000s was that there were extensive regulatory barriers in place to prohibit the shipping of wine across state lines—barriers the established wine industry loved because they protected them from upstart wineries. So one of the things we did at the FTC was to examine this empirically.

I remember having a conversation with C. Boyden Gray, who was an old friend of mine and had served as the White House counsel under President George H. W. Bush. He was representing the wine wholesalers, and they were dismayed that the FTC was looking to expand competition in wine.

One of the arguments they used was that these barriers somehow prevent underage drinking. I found that a curious argument; when I was a teenager, I didn't see a whole lot of demand for a really dry Chardonnay (online keg sales, however, would have been a different matter). Even assuming a big underage market for wine, the barriers on interstate shipment would have an impact only if potential underage drinkers somehow had a preference for out-of-state wine, as compared to in-state wine. That made no sense.

Nonetheless, we compiled evidence. We contacted various alcohol control boards and law enforcement officials in the states that allowed direct shipping of wine and asked if they'd seen any increase in drinking as a result of an increase in out-of-state wines. They responded, "None."

From contact lenses to coffins to wine, the pattern was always the same. The existing, powerful competitors lobbied their state regulators or state legislators to put up legal barriers to block anyone else from competing with them. The end result was, in the guise of helping consumers, such regulations tended to raise prices and hurt consumers.

A few years after our e-commerce task force began, it was particularly rewarding to see the Supreme Court strike down many of the barriers to the direct interstate shipment of wine. Repeatedly, the Court cited our task force's report.

———

I enjoyed the Federal Trade Commission, but it was a far cry from what I'd hoped for after the campaign and the long, hard work on the recount. I wanted to be the equivalent of Michael J. Fox's character in the movie *The American President*—the young, passionate idealist urging the president, in the heat of battle, to do the right thing.

I desperately wanted to be a real leader in the Bush administration—to have a senior post like so many of my campaign colleagues. When that didn't happen, and it became clear it wasn't going to happen, it was a crushing blow.

As a result, the first year of the Bush administration was one of the hardest of my life. But it also turned out to be one of the most important, because I couldn't blame anyone else for my situation—if I wanted things to change I had to look inside myself.

I think God knew what He was doing in 2001. At the tender age of thirty, I had already enjoyed a lot of success in my life, and I had foolishly come to believe in its inevitability. Campaigns are no place for candidates who think they walk on water.

In a grassroots campaign across a state like Texas, you talk to tens of thousands of voters in IHOPs and Denny's and VFW halls. You can come across as an arrogant little snot if you want to. But if you do, no one is going to vote for you.

Scripture tells us that pride cometh before the fall, and that lesson was taught to me forcefully. For the first time I had set a major life goal—getting a senior job in the administration—and I had worked hard for it. And I failed miserably.

With her usual insight, Heidi has observed that those two years changed me. Going through that experience altered my personality, and forced me to view the world differently, to treat others with greater respect and humility. It was a lesson I very much needed to learn.

In one of my favorite country songs, Garth Brooks sings, "Some of God's greatest gifts are unanswered prayers," and with me that certainly was true. If I had been appointed to a senior White House position, I would likely have become enmeshed in the many ill-fated decisions of the administration. And I certainly would have been overly impressed with myself, a failing that is all too common in Washington.

I needed to get my teeth kicked in. And if it hadn't happened, there's no way I would be in the U.S. Senate today.

CHAPTER 6

★

Upholding the Law

Well, what does he say?" asked Barry Goldwater.[1]

Dean Burch turned to the white-haired Arizonan who was his friend and mentor. "He says he hasn't been telling the truth."[2]

For the past five months of 1974, Burch had led the effort to protect the legal and political interests of Richard Milhous Nixon. Now Burch had to tell Goldwater what the senator likely already knew—the president of the United States was a liar.

Through the agonizing weeks and months as the Republicans had mustered a defense of the embattled White House, Nixon repeatedly had told his staff, his party, and his country that he had not participated in an illegal cover-up of an illegal robbery. Until that very morning, Burch had believed him.[3]

Burch arrived in Goldwater's office in the late afternoon of August 5 to bring the senator an advance copy of the president's latest statement about the two-year-old scandal. It addressed a tape of a conversation recorded in 1972, just a few days after five men were caught breaking into the Democratic National Committee's office at a then-little-known building complex called the Watergate.

Through his thick black horn-rimmed glasses so familiar that they had become his trademark, Senator Goldwater slowly read the document. A sense of sadness tinged with rage overcame him when he reached its eighth paragraph, a confession of what the secret tape would soon show: The President of the United States had personally directed an effort to frustrate a federal investigation and obstruct justice, and he had begun doing so almost as soon as he first heard the word *Watergate*.[4]

On the tape, which the Supreme Court had recently required the president to release to the public, Nixon had ordered that the FBI be instructed to stop investigating the Watergate break-in.

"Don't go further into this case," Nixon had barked into a hidden microphone, "period!"[5]

Like his former aide, Barry Goldwater had believed the president's assertions of innocence.[6] And it was Goldwater's opinion that mattered most. Indeed, in the decade since he'd lost the presidency in the Lyndon Johnson landslide, the conservative Republican had become an ornery, plainspoken, and grudging member of the GOP establishment that he had long railed against. Though he never liked Nixon that much, Goldwater had been among the president's most prominent and determined defenders. Now the senator had to decide what to do.

That night, Goldwater could not sleep.[7] His mind returned over and over again to the deception and dishonor of the president. His president. A president whom he had endorsed, campaigned for, and defended against a press corps that Goldwater would glare at and call a "rotten bunch" for their treatment of Richard Nixon.[8]

One of the options Goldwater considered on that long and sleepless summer night was to withdraw from his own reelection campaign, retire from politics, and make a televised plea for his party's leader to resign the presidency. If this kind of self-sacrifice would persuade Nixon to spare the nation the nightmare of an impeachment trial and restore confidence in our system of government, he would do it. But his wife talked him out of it. Little would be achieved, Margaret said, by Goldwater running away from Washington.[9]

The next day, the Republican Senate caucus gathered at noon for lunch with an unusual guest: Vice President Gerald Ford.[10] He had just come from a White House cabinet meeting, and he had more bad news. At the meeting, the president had vowed never to resign from office, even though the whole world now knew he was a crook after all. Nixon believed resignation would deliver a "hammer blow" to the office of the presidency.[11] It would be a desertion of "the principles which give our government legitimacy."[12] He was certain his refusal to resign was in the "best interests of the nation."[13]

When Ford finished speaking, a long, uncomfortable silence followed.[14] Goldwater and the Republican senators around him were in shock. Obstruction of justice and abuse of power were impeachable offenses, and in light of the smoking-gun tape proving Nixon's guilt, there was no doubt he would be impeached by the House of Representatives and convicted by the Senate. Was he really going to drag the country through the months it would take for the process to reach its inevitable conclusion?

"The best thing he can do for the country," an outraged Goldwater finally said to his colleagues, "is to get the hell out of the White House, and get out this afternoon!"[15]

The next day, Goldwater walked into the Oval Office to deliver that same message to President Nixon. With him were two other Republicans: House Minority Leader John Rhodes and Senate Minority Leader Hugh Scott. Goldwater felt the weight of history on his shoulders when he considered that it was the first time—and he hoped the last time—that the leadership of a president's political party called on the commander in chief to persuade him to resign from office.[16]

The president appeared surprisingly serene.[17] In the two decades since they had first served in Washington together, Goldwater had never seen the tightly wound Nixon so relaxed.[18] He was an enigma indeed.

Nixon leaned back with his feet on his desk, reminiscing about campaigns past, and treating his guests to a few monologues about Presidents Johnson and Eisenhower—history was on Nixon's mind. Then he casually said, "Well, we are all aware of why you are here. We might as well get down to it."[19]

Goldwater's two colleagues had asked him to be their spokesman, and although he had too much respect for the office of the presidency to tell Nixon to "get the hell out," he was determined to deliver the same message, with only slightly kinder words.

"Mr. President, this isn't pleasant," said Goldwater, "but you want to know the situation and it isn't good.[20]

"We've discussed the thing a lot and just about all of the guys have spoken up, and there aren't many who would support you if it comes to that," he said.[21] In other words, Nixon had little to no support—not even from his own party. They were unwilling to defend the indefensible.

"I kind of took a nose count today," continued Goldwater, referring to a future impeachment trial, "and I couldn't find more than four very firm votes, and those would be from older Southerners."[22]

He could have left it at that. The message that Nixon's presidency could not survive had been delivered, to the president's face, in the very house where the president had worked and slept for the past five and a half years. But Goldwater wasn't finished. He didn't just want his host to know he had lost others' support. The president needed to know he couldn't count on Goldwater, either. "Some are very worried about what's been going on, and are undecided," he said, "and I'm one of them."[23]

Goldwater was overcome with the sadness he had felt upon reading the statement Dean Burch had brought him two days earlier.[24] Beginning to tear up, he thought about how Nixon had been the most respected political leader on the face of the earth just two years ago when he was reelected in a sweeping landslide. Now he was ruined.[25] That night, due in no small part to the courage of members of his own party to speak the truth to a lawless president, Richard Nixon began writing his resignation speech.

On that day, the nation witnessed the central truth that is the strength of America's democracy: No one is above the law. Indeed, in a functioning republic, the rule of law is even more powerful than the president of the United States.

————

As phone calls go, it was one of the more frustrating of my life. Harriet Miers, the White House counsel for President Bush, was on one end of the line. Attorney General Greg Abbott of Texas and I were on the other. It wasn't long into the forty-five-minute call before we realized that much more separated us than the fifteen hundred miles between Washington and Austin.

We had called the White House counsel to argue against a seemingly improbable position that the Department of State was urging on the president: to side with the United Nations against his home state.

The first female president of the Dallas Bar Association and later the head of the Texas State Bar, Harriet Miers was a talented trailblazer for women and an accomplished commercial litigator. She's a friend. But her legal practice simply hadn't covered the structural questions that undergird our constitutional system.

And she was perfectly content with allowing the Bush administration to side with the United Nations against U.S. sovereignty. This was wrong, and it was inconsistent with what I knew to be the president's own conservative instincts.

Unfortunately, the issue was not being decided based on the Constitution. Instead it was being treated as a battle of personalities, an internal political matter. The year was 2005, the beginning of the second term of the Bush administration, and Condoleezza Rice and Alberto Gonzales were the newly appointed secretary of state and attorney general, respectively. In an effort to disprove the caricature of Bush as a "unilateral cowboy," State wanted the United States to give in to the World Court; Justice, quite rightly, wanted to defend American sovereignty.

Gonzales and Rice were headed for an Oval Office showdown— all but unprecedented for two aides so trusted by the president—but then they blinked. They reached a middle ground, and agreed on a compromise. And, for Miers, that was the end of the story.

Attorney General Abbott and I argued passionately that the president needed to be briefed on the profound consequences of ceding sovereignty to the United Nations, that it was unconstitutional and dangerous. But Miers was unpersuaded.

When we hung up the phone, Abbott and I simply shook our heads in dismay.

"Oh my God," he said.

"Oh my God," I agreed.

I was in the midst of litigating the biggest case of my tenure as solicitor general of Texas, and the lawyer with the ear of the president of United States had just made it clear that she was in no position to persuade him that his State Department was about to put him, and the federal government, on the wrong side of history, the Constitution, and the state of Texas—and increasingly, it was sad to say, on the wrong side of his own conservative principles.

In November 2002, a few weeks after Texans elected Greg Abbott to succeed John Cornyn (now my Senate colleague) as attorney general, I received a call from an old friend. "Greg Abbott is looking for a solicitor general," he said. "Are you willing to have your name considered?"

The solicitor general is the chief lawyer for Texas before the U.S. Supreme Court and all state and federal appellate courts. Starting two or three decades ago, state attorneys general began creating solicitor general offices, modeled after the federal office, which represents the executive branch before the Supreme Court. Their goal was to improve the caliber of appellate advocacy for the states, which had often been outmatched by private advocates who had spent lifetimes honing their appellate skills.

"Give me a day to think about it," I told my friend. On one hand, the job would undoubtedly provide fascinating and important work. On the other hand, it would mean leaving a position I liked. Working at the Federal Trade Commission was not my first choice in the Bush administration, but I'd come to enjoy it. To further complicate the question, Heidi had moved on from the U.S. Trade Representative's Office to the Treasury Department, where she was very much enjoying running the Latin America office.

"Let me talk to Heidi," I said, "and I'll call you back tomorrow."

At the time, Texas had one of the strongest state solicitor general's offices in the country. That tradition began with John Cornyn, who, as attorney general, had created the new office and brought on a former U.S. Supreme Court clerk named Greg Coleman to head it. Coleman hired a cadre of very talented and experienced lawyers, and many of them have since spent their careers serving Texas in exemplary fashion.

When I talked with Heidi that night and asked her what she thought about the possibility of my becoming solicitor general, she told me I should definitely go for it. It was only afterward that she admitted she thought I didn't have a prayer of getting the job, and so it had been easy for her to encourage me to toss my name in the hat.

Greg Abbott is a remarkable individual. Just after he had graduated from law school and taken the bar exam, Abbott was out running when a tree that had been struck by lightning fell on him and broke his back. He was in his twenties when he found himself paralyzed from the waist down.

Abbott spent over a year in rehabilitation dealing with the accident, but he also got on with his law career. After practicing law for several years, he became a respected state trial judge, and then a state Supreme Court justice. In 2002 he was elected attorney general, and he brought the attitude of a skilled jurist to an office that in far too many states is overly politicized.

Even after my interview with Abbott went well, I wasn't optimistic about my chances. Attorney General Abbott, who was then in his mid-forties, had been frank in our meeting that he had some real concerns about my youth and inexperience in the courtroom. Those concerns were understandable. I was thirty-one years old, and I had only ever argued two cases in court—and none in front of the Supreme Court. If he hired me, I would be the nation's youngest state solicitor general.

A couple of weeks after our meeting, in the early morning a few days before Christmas, he called me and said the job was mine if I wanted it. I was astonished, but not nearly as much as Heidi.

Heidi and I talked about it and prayed about it. It would mean weekly trips for her and me back and forth between Austin and Washington, where Heidi would continue to work. That would be tough for newlyweds. But like the frontier pioneers who chalked "GTT" on their front doors when they took off in search of new opportunities, in February 2003, I was gone (back) to Texas.

———

The most profound privilege of serving as solicitor general is representing Texas before the U.S. Supreme Court. Litigating a case before the Supreme Court is, as one might imagine, a unique experience. I say this having had the privilege of seeing it from both sides—as a clerk for the Chief and then as a litigator myself.

My first case before the Court is so etched in my memory that I can remember the exact date, October 7, 2003. I knew Chief Justice Rehnquist well enough to know that he would show his former law clerk no mercy. Neither would any of the other justices whom I had come to know during my clerkship. I had to be on top of my game—in fact, I felt a special burden. I had such deep respect for the justices that there was no way I could allow myself to disappoint them.

I prepared for my appearance for two months, and practiced for the oral argument with four grueling moot courts. Indeed, from the beginning of August until the night before the hearing, I did almost nothing other than work on the case. And all along I knew that we didn't really have a chance of winning.

The case was *Frew v. Hawkins*. In it, lawyers representing a class of one million Texans accused the state of not complying with federal Medicaid requirements. At stake in *Frew* were hundreds of millions of dollars in taxpayer money and the question of whether an unelected federal judge can override budgetary decisions made by a state's legislators. We argued that budgetary decisions should always be in the hands of the people's representatives. But our opponents had a strong counterargument: It was state officials who had assented to the settlement agreement in the first place. A prior attorney general, Democrat Dan Morales, who later served prison time for corruption, had explicitly agreed to the consent decree. The court was merely enforcing an agreement to which the state had already affirmatively consented.

Even though I had serious doubts about our chances of winning, I felt ready for the oral argument—confident, even. I had prepared diligently, and I knew the case backward and forward. But once I went to bed, I didn't sleep a wink the entire night. My stomach quivered, my mind raced. I kept staring at the ceiling, considering questions and answers over and over in my head. Perhaps my subconscious was telling me I wasn't quite as confident as I had thought.

Lawyers tend to be superstitious. I had a favorite pair of boots—my lucky black ostrich "argument boots"—and I had worn them in every single argument I had given as solicitor general. But I hesitated before putting them on that morning.

Rehnquist was a stickler for wardrobe. Indeed, he'd previously reprimanded a lawyer for wearing a brown suit. "That is not appropriate attire for this courtroom," the Chief told him. And there was an excellent chance that if he saw my boots, he would say the same thing to me.

Lacking the courage of my convictions, I'm ashamed to say that I went to my closet, put away my boots, and instead wiped away the years of dust on an old pair of wingtips.* I was already starting on an uncomfortable note.

One of the first things I saw as I made my way to the lectern was

---

*When John Roberts became Chief Justice, I finally found my courage. Shortly after he was sworn in, I asked him, "Tell me, Mr. Chief Justice, do you have any views on the appropriateness of boots as footwear in oral argument?" With a grin, Roberts replied, "Ted, when representing the state of Texas, they are not only appropriate, but required." From that point on, in every argument I made before the Court, I wore my boots.

the familiar visage of Chief Justice Rehnquist in his flowing black robe with four gold stripes on either sleeve. I stood at the podium, breathed in deeply, and began with the customary "Mr. Chief Justice, and may it please the Court."

Standing there, one of the first things that struck me was just how small the courtroom was. Although majestic—with an ornate ceiling inlaid with 24-carat gold—the courtroom itself is smaller than a typical trial court. Maybe it says something humbling about our democracy that the most important court in the land is not some grand chamber, but a room about the size of a small basketball court. In fact, directly above it is an *actual* basketball court, which is nicknamed internally as "the highest court in the land."

In such a small environment, I stood all of about seven feet from the justices, so close I could almost shake the hands of nine of the smartest and most dangerous legal minds in the United States.

Many people think of TV shows like *Perry Mason* or *Law & Order* when they imagine a courtroom argument, but proceedings in the Supreme Court are quite different from those of lower courts. There are no witnesses. There is no evidence. There are no grand orations. Instead, Supreme Court arguments consist of relentless questioning from the justices in cases about which they are already expert. In typical arguments, advocates will utter just a few sentences before the justices begin firing hostile questions at them. In a contentious case, for the rest of the argument, the blitz of questions doesn't cease.

For my thirty minutes in *Frew*, there was not a single friendly question directed toward me. The justices were ripping me limb from limb. I felt like a chunk of tuna thrown to a school of sharks. For a long time, my answer to questions were a combination of "yeah . . . wha . . . but . . ."

Many oral arguments are lost on a single question that the justices ask repeatedly (with different variations) and for which there is simply no good answer. In *Frew*, that question was some version of "How can you say the state didn't consent to a *consent* decree?"

By the time Chief Justice Rehnquist closed the session—"Thank you," he said in his deep baritone, "the case is submitted"—it was pretty clear the Court was going to rule against me.

And, just a few months later, it did. Unanimously, 9–0. After the case was decided, I visited with the Chief Justice in his chambers over tea.

"Well, they say that with your first argument, you should pick a case you can't lose or you can't win," he said with a smile. "Ted . . . I think you chose wisely."

———

Although our defeat in *Frew* was discouraging (but not unexpected), over time we ended up winning repeatedly before the Court. And the nature of the legal issue in the case—an unsettled question of constitutional law—foreshadowed the kinds of matters that would define my five and a half years in Austin. During that time, Texas found itself at the forefront of one major constitutional battle after another.

Some of that was good fortune, but some of it we affirmatively sought out. When Attorney General Abbott appointed me to the job, he told me, "I want you to look across the country, and if we can step up, defend conservative principles, and make a meaningful difference, go do it." That was an amazing mandate. I knew then that coming back to Texas to serve under him had been the right decision.

The thing that amazed me the most about Greg Abbott was his character. He is a man who courageously fights for his principles and who generously shares the credit with the team around him. During my tenure as solicitor general, he allowed me the opportunity to tackle high-profile issues that few other statewide-elected officials would have given to a subordinate. That was a reflection of his humility and deep confidence.

And he is a man who managed to reject the natural inclination to become bitter or angry with God over being confined to a wheelchair as a young man. I have often wondered how I would have reacted to such an injury. I have no certainty that I would manage his equanimity. But Greg Abbott is at peace with himself and his disability, and Texas made a wise decision in 2014 when it elected him as its governor. I'm confident he'll prove to be an extraordinary one.

One of the earliest opportunities to fight for conservative principles presented itself just a week after the oral argument in *Frew*. On October 14, 2003, the Supreme Court agreed to review a decision by a federal court in California holding that the Pledge of Allegiance cannot be recited in public schools. The appellate court concluded that, because the pledge says "one nation, *under God*," it violates the First Amendment's proscription against laws "respecting an establishment of religion."

It was no surprise that the decision came from California. The Ninth Circuit—in which the California federal courts are a major component—tended to issue liberal, sometimes extremely liberal, opinions. But the decision was deeply misguided. It was typical of those on the left who are intent on eradicating any vestige of religion from the public sphere. The First Amendment was not adopted to create government hostility to religion; rather, the First Amendment exists to protect the religious liberty of every American.

Texas proudly took the lead in defending the Pledge of Allegiance. My team and I wrote an amicus brief, which is a brief filed by someone who has an interest in a case but is not a party to it. We argued, "From the time of the Founding, our Nation has recognized her religious heritage, and the Constitution has never been understood to prohibit those acknowledgments. After all, the national motto is 'In God We Trust.' Our Declaration of Independence refers to rights 'endowed by our Creator.'" Our brief pointed out that the Supreme Court even "begins its own proceedings with the cry, 'God save the United States and this Honorable Court.'"

As the deadline for submitting the brief approached, I spent several days on the phone calling solicitors general in other states to urge them to join our brief. Most had joined, but a handful refused. On the final day, I called each of the holdouts back. "I just wanted to check again and see if your boss wants to join this brief," I said. "If not, fine, I respect that your boss wants to be one of the few attorneys general in the country who doesn't support the Pledge of Allegiance. That's a real profile in courage."

In the end, no attorney general wanted to be left standing without a chair when the music stopped. All fifty attorneys general signed Texas's brief, the first time to my knowledge that every state has signed a single brief submitted to the Supreme Court. And five months later, the Court unanimously reversed the decision of the lower court. We won, and children in California schools were once again free to pledge their allegiance to "one nation, under God." *

――――――

The question of the Pledge of Allegiance's constitutionality was one whose answer I felt pretty confident about. But in my first year as solicitor general, the attorney general called with another question I had never considered.

――――――

* The Court vacated the lower court's decision on procedural grounds, so it didn't address the underlying merits. But the result of the decision was that students were once again free to recite the Pledge of Allegiance.

"Ted," Abbott asked, "can the Speaker of the Texas House arrest fleeing representatives?"

Abbott was referring to a group of Democratic state representatives who had fled the state in order to prevent a legislative quorum, which would have allowed the Republican majority to enact a new electoral map of congressional districts.

Every ten years, per the U.S. Constitution, the nation conducts a census. State legislatures then redraw congressional maps to reflect the number of congressional seats that a state's new population entitles it to. In the 1980s and '90s, Democrats controlled the state legislature, and they used their power to gerrymander the map to favor Democrats. As a result, in 2003, 53 percent of our congressmen were Democrats, even though only 44 percent of Texans voted for Democratic congressional candidates.

By 2003, Republicans had been elected to majorities in the Texas Legislature, and they decided to draw a new map that reflected the views of Texas voters. What entailed was an epic war waged on political and legal battlefields, including renegade Democrats trying to flee the state to prevent a vote. Research into the Texas Constitution revealed the answer to Abbott's question was "yes." The Speaker has full authority to arrest absent legislators; indeed, the U.S. Constitution has the same language, and in the eighteenth century, leaders of Congress actually had put members in leg irons to secure their attendance.

Ultimately, after our office prevailed in several early legal skirmishes, the fleeing legislators returned and the Texas Legislature passed a redistricting map that redrew the geographic boundaries of federal legislative seats. It then became my responsibility to defend the legality of that map in a federal trial before a three-judge court. The central issue was who has the authority to determine the shape

of a state's congressional districts—the state's elected officials or unelected federal judges. To me, at least, the answer seemed obvious.

One of the more unpleasant aspects of that litigation was that
redistricting cases are among the most expressly racial cases that
ever make their way into a twenty-first-century courtroom. Normal
Americans don't treat race as a dominant criterion. We don't think
of people as "my black friends" or as "my Hispanic coworkers." A
friend is simply a friend. A colleague is simply a colleague.

But congressional seats are different—at least when it comes to
redistricting litigation. Seats are routinely divvied up by the litigants
on both sides and by the courts in terms of "African-American
seats," "Hispanic seats," and "Anglo seats." The precedents in redistricting litigation have made electoral maps' effect on racial groups
the decisive issue in a court of law. As Chief Justice Roberts wrote
in the context of affirmative action and education, "It is a sordid
business, this divvying us up by race."*

Because of the inherently racial nature of this business, litigating redistricting cases is a fate I wouldn't wish on my worst enemy.

---

*John Roberts was appointed Chief Justice by President Bush in 2005. Roberts was the
only Supreme Court clerk ever to succeed his boss to the Court. This was fitting, since
no one was more intellectually similar to Rehnquist than John Roberts. Before becoming
Chief Justice, Roberts was an extraordinary advocate; indeed, when I began practicing,
I worked hard to try to emulate his argument style. In his oral arguments, he never got
emotional. He simply answered each question that was asked of him. This sounds easy and
obvious, but in fact it is exceedingly difficult to resist the temptation to dodge hard questions or to answer the question you wanted, not the one you actually got. You do not get
to be the best Supreme Court advocate of your generation, as John was, without having an
exquisite understanding of what makes each of the sitting justices tick. With his immense
credibility before the Court, Roberts had consistently managed to get the swing votes to
sway in his direction. Therefore, whenever I appeared before the Roberts Court, I would
always listen carefully to his line of questioning. Often he would be raising arguments that
he deemed most likely to win over swing votes like Justice Kennedy. And whatever line he
was posing, I would try to follow.

But despite my distaste for certain aspects of the litigation, I spent the Christmas season of 2004 in federal court helping lead the trial team in defense of my state's electoral map.

One exchange in that trial was particularly memorable—and perhaps even more absurd than the state of the law in this area. It involved state representative Ron Wilson, a flamboyant African-American Democrat with an orange Lamborghini and a penchant for black leather trench coats.

Wilson represented the inner city of Houston, which had a large percentage of African-American constituents. Because he was a Democrat, Wilson was expected to oppose the Republicans' redistricting map. But Wilson realized that although the map increased the number of districts with Republican majorities, it also increased the number of districts with African-American and Hispanic majorities. The losers were white Democratic politicians, who would no longer benefit from the gerrymandering that had insulated them from Texas's conservative voters.

Representative Wilson decided to defy Democratic colleagues and support the redistricting plan. "It was never a question of if the redistricting bill would pass," he said. From his perspective, the question was "Do you stand on the railroad track and try to stop the train, or do you try to get some of your people on the train and not get run over?" [26]

This decision did not endear Wilson to a number of his fellow Democrats, including Lee Godfrey, one of the opposing lawyers in the redistricting litigation. During Godfrey's cross-examination of Wilson at trial, Godfrey accusatorially observed that Wilson comprised "one hundred percent of the African-American legislators" in support of the new map.

Wilson, never one to back down from a fight, shot back, "I am the only one who had the 'things' big enough to do it." He gestured accordingly.

I have to confess I was stunned. It is not every day that a witness refers to his genitals in federal court.

Even more unusual is an attorney who takes the bait.

"I presume the 'things' you refer to are not visible?" said Godfrey.

"You want to see them?" demanded Wilson.[27]

Godfrey paused, ready to accept. But the judge wisely interrupted, "Move on. Move on." It is a sordid business, this redistricting litigation. But not that sordid.

As Sun Tsu famously said, "Every battle is won before it is ever fought." He meant in part that battles are won by choosing the terrain on which the fighting will occur. Litigation is no different. Nothing is more important than how you frame the narrative, which is the intellectual terrain on which the legal battle will be fought.

In every case, I tried to think about what the judge would tell his or her kindergarten-aged grandchild when the child asked, "Pawpaw, what did you do today?" If you can successfully frame the judge's one-sentence answer to that question, you're probably going to win the case. If you can't, you're going to lose.

When the redistricting litigation arrived at the U.S. Supreme Court—after we won in the district court—the opposing side argued that it was unconstitutional for a legislature to be overly political in drawing a redistricting map. They wanted the narrative to be that the mapmakers were hyperpartisan.

We could have tried to argue that redistricting in Texas had been nonpartisan. The problem was that any such argument would have been ludicrous. Politics permeated every facet of the redistricting process, and I was not about to argue otherwise. Even attempting such an argument would have sacrificed my credibility with the

Court, and once an advocate has lost his credibility, he has lost his case.

Instead, I tried to reframe the narrative. "The central issue," I said, "is determining which institution is constitutionally vested with the primary responsibility for redistricting: elected legislatures or federal courts." That was what I wanted Justice Kennedy, the likely swing vote, to tell his grandchild when he was asked what that day's case was about.

I emphasized in my argument that the framers of the Constitution understood politics. They knew full well that when you give redistricting decisions to elected politicians, they will make political decisions. In fact, the term *gerrymandering* comes from Elbridge Gerry, a delegate to the Constitutional Convention whose Massachusetts congressional district was so convoluted it looked like a salamander on the map.

Framers like Gerry deliberately entrusted elected state politicians with defining the geographic boundaries of congressional districts. Doing so protects our constitutional structure to keep that power in the hands of elected legislatures, because if legislators overstep, voters have the ability to "throw the bums out." In contrast, voters have no control over the redistricting process if power is taken from the people's representatives and given to unelected federal judges.

Ultimately, our argument prevailed. In an opinion authored by Justice Kennedy, the Court declared, "We reject the statewide challenge to Texas' redistricting as an unconstitutional political gerrymander. . . ."* Justice Kennedy wrote that our opponents "established no legally impermissible use of political classifications." In

---

*The challengers did prevail with regard to the legality of the drawing of one particular congressional district.

other words, of course the legislators who had redrawn Texas's congressional districts had been political, but there was nothing unconstitutional or illegal about politicians being political.

The Court's decision had lasting consequences for Texas. In the next election, Democratic candidates for Congress received the same percentage of statewide votes as they had in the election before the legislature's redistricting. But in the earlier election, Democratic candidates had won an overrepresentative 53 percent of the state's congressional seats. In the subsequent election, they won only 41 percent of them.[28]

Depending on your political perspective, it is debatable whether the Texas congressional delegation now consists of better representatives for the state. But it is indisputable that the Texas delegation is now more representative *of the actual voters* in the state. I do not think it is a stretch to suggest that that is how a system of self-government is supposed to work.

———

If framing the narrative and guarding one's credibility are the first two rules of litigation, a third rule—perhaps the most obvious sounding of them all—is: Don't appeal a victory. But in the course of defending a public display of the Ten Commandments, I confess to having flagrantly violated that rule.

The case arose when an atheist—a homeless man and former lawyer—was walking through part of the twenty-two acres surrounding the Texas State Capitol. Displayed among sixteen other monuments and twenty-one historical markers is a stone monolith inscribed with the Ten Commandments. The atheist took offense at what he believed was a promotion of religion by the state, and in 2003 I found myself arguing against his claim in the Fifth Circuit Court of Appeals. As with the Pledge of Allegiance case, the Ten

Commandments case was an important opportunity to push back against those who want freedom of religion to be the absence of religion. I believe strongly the opposite is the case, and that when the Founders enshrined religious liberty they were welcoming all Americans to worship openly and as they choose, not trying to force them to hide or deny their faith.

We are a nation founded by men and women who were fleeing religious persecution and coming to a land to seek out and worship their Lord, according to the dictates of their own conscience, without government getting in the way. On that understanding, the purpose of the Establishment Clause was to prevent the federal government from coercing individuals to engage in any particular religious belief or practice. It was not—and is not—about mandating government hostility to religion.

Indeed, the acknowledgment of our shared Judeo-Christian heritage has been ubiquitous in the history of our country—as we had argued in the Pledge of Allegiance case. Among the best representations of that heritage is the display of the Ten Commandments in the courtroom of the Supreme Court itself—where the tablets appear no less than forty-three times.

In the course of that argument before the court of appeals, I noted that the precise language on the Texas monument had been composed by a priest, a minister, and a rabbi. At that point, one of the federal judges on the panel observed, "That sounds like the opening of a joke."

I rather foolishly responded, "Yes, your honor . . . but I'm pretty sure no one walked into a bar."

Fortunately, even though I had violated another cardinal rule of litigation—do not attempt humor because of the risk of a monumental backfire—the judges took pity on me. They laughed heartily.

Even better, they unanimously ruled in our favor. The court held that Texas's public display of the Ten Commandments did not violate the First Amendment's prohibition against laws "respecting an establishment of religion."

Ordinarily, when you have won in the court of appeals, you try very hard to convince the Supreme Court *not* to hear your case, because if the justices decline to hear it, the lower court decision stands and you win. But across the country, other appellate courts had found public displays of the Ten Commandments to be unconstitutional. Abbott and I had a long discussion about the strong likelihood that the Supreme Court would take one of these cases in the next couple of years. So we asked ourselves, "Which case has the best chance of protecting the Ten Commandments, and the freedom of religion?"

We concluded that no state had a better chance of securing a victory on this issue in the Supreme Court than Texas did. Other states had to overcome ill-advised statements by government officials that the explicit purpose of their displays of the Ten Commandments was religious. Not so in Texas, where the Fraternal Order of Eagles—a secular service organization—had erected the monument to promote morals and responsibility among young people. Moreover, the statue had been standing since 1961 and was surrounded by dozens of secular monuments. The Supreme Court would consider both of those factors—being part of a historical tradition and being associated with nonreligious displays—to weigh in favor of a display's constitutionality.

We therefore took the unusual step of filing a brief before the U.S. Supreme Court saying that if the Court was inclined to consider a case raising this legal question, it should agree with our *opponent's* request that the Court hear our case. That was a high-risk proposition, because the Supreme Court is extremely unpredictable on cases involving religion. Only three of the nine justices were sure to rule in our favor. If we couldn't persuade more than Chief Justice Rehnquist and Justices Scalia and Thomas, we would lose the entire case.

Our gamble began to pay off when the Court agreed to hear two cases about the Ten Commandments, both at the same time. One arose in Kentucky, where a host of factors made victory for the state less likely. The other was ours.

For this particular case, I recommended to Attorney General Abbott that he present the oral argument himself. He had not yet argued in front of the Supreme Court, and I told him, "If you're going to argue a case, this is the right one for you to choose. The issue is incredibly important, and I believe we can win." He agreed, spent two months holed up in his office preparing, and did a superb job. At the end of the oral argument, the presiding justice—Justice Stevens, rather than Chief Justice Rehnquist, who was ill—did something that I had never seen in any argument before and that I have not seen since: He made a point of complimenting Attorney General Abbott for his particularly excellent oral advocacy.

"General Abbott," said Justice Stevens to the wheelchair-bound attorney general, "I want to thank you for your argument and also for demonstrating that it's not necessary to stand at the lectern in order to do a fine job."

Typically, the biggest cases at the Supreme Court are decided in June, the term's last month. In 2005, the first twenty-six days of June came and went with no decision in our Ten Commandments case. Finally, on the Court's last day, June 27, Chief Justice Rehnquist said from his seat at the center of the Court's bench, "I have the opinion of the court to announce in Van Orden against Perry."

His voice was weak and gravelly to the point of being almost unrecognizable, a casualty of the cancer that would take his life sixty-eight days later. He could only speak five or six words without pausing and taking a deep breath. But it was important to him to sit on the bench that day and announce his decision in person.

"The Court of Appeals for the Fifth Circuit affirmed the district court's finding that the monument does not contravene the First Amendment's Establishment Clause," he said. "The judgment of the court of appeals is affirmed."

With those words, the Chief Justice completed a jurisprudential arc that began with his early years on the bench in the 1970s and early 1980s. Back then, as I mentioned earlier, he dissented so frequently that he had earned the nickname the Lone Ranger. Rehnquist, for example, had written a dissent in a 1980 case in which the Court held that displaying the Ten Commandments in a public school is unconstitutional. In that earlier case, he wrote, "The Establishment Clause does not require that the public sector be insulated from all things which may have a religious significance or origin. This Court has recognized that 'religion has been closely identified with our history and government,' and that '[t]he history of man is inseparable from the history of religion.'"

Twenty-five years later, Chief Justice Rehnquist's opinion in *Van Orden* made the same argument, sometimes with the exact same words. This time, however, he was writing for the Court's plurality.

After announcing the Court's decision in *Van Orden*, the Chief Justice said that "Justices Scalia and Thomas have filed concurring opinions. Justice Breyer has filed an opinion concurring in the judgment. Justice Stevens has filed a dissenting opinion in which Justice Ginsburg has joined. Justice O'Connor has filed a dissenting opinion. Justice Souter has filed a dissenting opinion in which Justices Stevens and Ginsburg have joined." He then added, "I did not know we had that many people on our Court."

The courtroom erupted in laughter. On an issue that evoked strong emotions and that had fractured the Court into six opinions, Chief Justice Rehnquist used humor to bring everyone together. It was a graceful and fitting conclusion for the career of a man who had never wavered in the defense of his principles, but who had always shown kindness and respect to those who disagreed with him.

From both a professional and personal perspective, two elements of the case were particularly rewarding. First, *Van Orden* vindicated the decision that Attorney General Abbott and I made to encourage the Supreme Court to hear our case, even though we had prevailed in the lower courts. On the same day *Van Orden* was decided, the Court struck down Kentucky's display of the Ten Commandments. Its fact pattern presented the state with far more legal obstacles than Texas faced. If the Supreme Court had not heard our case as well, it would have been a dark day for religious liberty.

On a personal level, it meant a great deal to be part of the team litigating a case that allowed Chief Justice Rehnquist to uphold the principles that he had espoused his entire life. He was not just my former boss. He was also my close friend. Our case was the last decision that William Rehnquist ever authored. Just two months later, I joined his other law clerks as pallbearers at his funeral. He was ably succeeded by one of those former clerks, John G. Roberts, but our country had lost one of its greatest jurists.

By any measure the biggest case of my tenure as solicitor general was *Medellín v. Texas*—the case that led to our phone conversation with White House Counsel Harriet Miers.

Medellín began in 1993 with two innocent girls, fourteen-year-old Jennifer Ertman and sixteen-year-old Elizabeth Pena. The teenagers were walking home in Houston, taking a shortcut through a secluded area in order to make it home by their curfew. They ran into a group of about half a dozen members of the "Black and White" street gang. Elizabeth was grabbed. Jennifer escaped, but when she heard her friend's screams, she returned to try to help.

For the next hour, Elizabeth and Jennifer were repeatedly raped. Then they were murdered.

Among the assailants was an illegal immigrant named José Ernesto Medellín.

After Medellín's arrest, he wrote a four-page, handwritten confession. It may be the most horrifying document I have ever read. In it, a cold, callous Medellín dispassionately described how the gang summarily decided to end Jennifer's and Elizabeth's lives. How he joined in their rape. How they stomped on the neck of one girl and strangled her to death with a belt. How they strangled the other girl with a shoelace, Medellín holding one end, his friend the other as they cut into her throat.

The most chilling detail, which will always remain with me: The youngest girl was wearing a Mickey Mouse watch, which he proudly kept as a souvenir.[29]

Not surprising, José Medellín was convicted and sentenced to death.

A decade later, while Medellín was on death row, the case took an unusual turn. In 2004, the International Court of Justice—the judicial arm of the United Nations, also known as the World Court—issued an unprecedented decision. It ordered the United States to reopen the murder convictions of Medellín and fifty other Mexican nationals. Never before had any international court attempted to directly bind American courts, much less set aside final criminal convictions.

The World Court based its decision on the Vienna Convention, an international treaty that the United States ratified in 1969. The court concluded that the Vienna Convention requires the United

States to inform foreigners of their right to contact their consulate if they are arrested in the United States, even if they, like Medellín, have lived in the United States for almost their entire lives. Because Medellín had not been informed of this right by the police who arrested and questioned him, the World Court ordered the United States to provide Medellín with "review and reconsideration" of his conviction and death sentence.

Under U.S. law, the Vienna Convention did not entitle José Medellín to a new trial. Pursuant to ordinary rules of criminal procedure, Medellín had forfeited his right to invoke the Vienna Convention when his lawyers failed to raise the issue in his trial and original appeal. And, more important, only Congress can make the Vienna Convention binding domestic law that is enforceable in American courts, and Congress had chosen not to do so.

None of this mattered to the World Court. It didn't respect the prerogatives of Congress. It didn't respect the rules of Texas's criminal courts. And it certainly didn't respect the sovereignty of the United States of America.

Unfortunately, as already mentioned, Secretary of State Condoleezza Rice wanted us to accede to the World Court. Condi is someone I've known since the Bush campaign; she's a friend, and she was Heidi's boss at the National Security Council in the White House. (In 2004, Heidi had moved from Treasury to the NSC, to serve as the economic director for the Western Hemisphere.) I very much respected Rice's experience and judgment, but on this question I believe she was wrong.

The State Department wanted President Bush to oppose the final judgment of the Texas criminal justice system, to support the decision of the World Court, and to subject state and federal courts to the authority of the judicial arm of the United Nations, even though every previous president—including Jimmy Carter and

Bill Clinton—had rightly taken the position that courts outside the United States cannot bind American courts.

The decision to change course and try to subject Texas courts to the United Nations was in keeping with the prevailing mood in parts of the administration at the beginning of President Bush's second term: The president had taken a public relations beating for allegedly alienating allies with the Iraq War, and over the objections of hard-liners like Dick Cheney and Donald Rumsfeld, many high-ranking officials appeared determined to make the next four years a "We Love the World" term.

Therefore, several weeks after our unproductive phone call with White House Counsel Harriet Miers, I received a call from President Bush's solicitor general, Paul Clement.

"Ted," he said ominously, "are you sitting down?"

"Paul, that's not a good way to begin a conversation," I replied, sort of laughing. "I'm worried what's coming next."

In his late thirties at the time, with thinning hair and a shy smile, Clement is the mild-mannered native of a small town in Wisconsin. He is also one of the most talented appellate advocates of his generation. Now in private practice, Clement is hired to argue more Supreme Court cases in a single year than most appellate experts will argue in a lifetime.

When Clement called me in early 2005, he knew how strongly I felt about the constitutional issues in the *Medellín* case, and he didn't relish being the bearer of bad news. But he was a good soldier, even in an administration that didn't always stand up for its conservative principles. So he informed me that President Bush had decided to sign a two-paragraph order purporting to compel the state courts to obey the World Court.

Clement explained that the administration admitted that no international treaty *required* state courts to obey the World Court. But they were taking the position that the president's inherent authority as commander in chief empowered him to promote "international comity"—goodwill among nations—by forcing the courts of Texas to obey the World Court. This approach, he continued, would allow the president—at his sole discretion—to turn on and off the power of the World Court over legal proceedings in the United States.

My reaction was, to say the least, unenthusiastic.

"Ted," said Paul, doing his best to put a good spin on this surrender of American sovereignty and blow to federalism and the separation of powers, "you should be happy with this, because the great thing is that with this new power we're claiming, the president keeps his finger on the trigger. He gets to decide when to employ it."

"I have two thoughts," I said calmly. "Number one, that's not very comforting, given how it's being employed right here and now." I was thinking about all the crackpot dictators and America-haters who wield outsize influence at the United Nations—as well as the memory of two raped and murdered teenage girls who might now never have justice.

"But number two, Paul, as Scripture says, 'There came a Pharaoh who knew not Joseph and his children.' George W. Bush is not going to be the last president of the United States, and if he has this power, what about the next president, or the next president, or the next president?"

Paul had no good answer.

The president's executive order—written as a memorandum to his attorney general—left the state of Texas with a difficult choice. George W. Bush was the former governor of Texas. He was a Republican. He was a friend of Greg Abbott's and mine, not to mention my former boss. Would we really argue in court that his executive order was the abuse of power we believed it to be?

I am proud to say that Abbott never wavered. He knew that

Texas was right and Bush was wrong, and he decided to stand up for the principle that no president—not even a friend and fellow Republican—can defy the Constitution. And so I went to the Supreme Court and argued on behalf of Texas that, with or without an executive order endorsing it, the United Nations' World Court has no authority to bind the courts of the United States.

Virtually every academic and media observer predicted we were toast. They believed there was no argument Texas could make that would carry the votes of five justices on the Supreme Court, because the narrative proposed by the opposing side appeared so compelling: Texas cannot flout international commitments, international courts, and the president of the United States. According to their logic, Justice Kennedy would come home and tell his grandchild, "Today, we had a case about whether a rogue state can defy treaties ratified by the United States of America."

If that narrative had prevailed, our chorus of media and academic critics would have been right. We would have lost. So we had to change the narrative.

Many states would have litigated *Medellín* as a federalism case. They would have focused on the argument that the federal government doesn't have the authority to set aside a state's rules of criminal procedure. But the problem with that strategy is that it played right into the narrative that Texas was defying the treaty obligations of the federal government.

For that reason, we made a different strategic decision, which was to make *Medellín* into a separation of powers case. The lead argument we presented in our brief was that the president's order violated the authority of Congress, because the Senate ratified the Vienna Convention on the explicit understanding that the treaty was *not* binding on state and federal courts. Our narrative was that the president had usurped Congress's power by executive fiat.

Our second argument was that the president's order usurped the authority of the Supreme Court, because the Court, in a case called *Sanchez-Llamas*, had recently adjudicated a related question about whether the World Court could bind the American justice system. In effect, the president's order said it didn't matter how the Supreme Court interpreted federal law; what mattered was how the man in the Oval Office interpreted federal law. That, of course, is not what Chief Justice John Marshall wrote in the landmark 1803 Supreme Court decision, *Marbury v. Madison.*

It was one of the great privileges of my life that I had the opportunity to stand before the Supreme Court and rely, as a principal authority, upon *Marbury v. Madison*, which held, "It is emphatically the province and duty of the judicial department to say what the law is."

Finally, as a third argument, and very much a tertiary argument, we raised the federalism concerns—that the president's order violated the authority of the states. Neither the president nor Congress may, as our brief said, "alter the structure of the state government or commandeer the state judiciary in order to implement federal policy." I believed this argument was valid, but I didn't have confidence that it alone would be enough to attract the support of five justices. As a result, in a fifty-page brief, we gave it only about four pages.

Our approach to the case began to bear fruit at the oral argument on October 10, 2007. During Solicitor General Paul Clement's time at the podium in defense of President Bush's executive order, Justice Scalia said to his former law clerk, "Usually, when we have treaties that are not self-enforcing, the judgment of whether that international-law obligation shall be made domestic law is a judgment for the Congress."

Quite so. Justice Scalia's observation perfectly captured our separation of powers narrative, as did his subsequent ridiculing of

Clement's position. "You're telling us that, well, we don't need the Congress," said Scalia. "The president can make a domestic law by writing a memo to his attorney general."

After Justice Scalia captured the essence of our first separation of powers argument (that the president was usurping Congress's authority), Chief Justice Roberts captured the essence of our second separation of powers argument (that the president was usurping the Supreme Court's authority). The Chief Justice asked Clement if the president "can take action that's inconsistent with the determination of federal law by this Court?" When Clement agreed that the president cannot, Roberts replied, "I thought we determined in *Sanchez-Llamas* that the treaty did not mean what the [World Court] said it means in this case."

Minutes later, Justice Kennedy echoed this theme. "I agree that we should give [the World Court's] determination great weight, but that's something quite different from saying that [the president] can displace the authority of this Court on that issue of law."

At that moment, before my own argument had even begun, I breathed a sigh of relief. If we had Justice Kennedy, we probably had five votes.

In the end, when the decision was handed down five months later, we didn't have five votes . . . we had six.

With an opinion written by Chief Justice Roberts and joined by the Court's conservatives and also the liberal Justice John Paul Stevens, the Supreme Court agreed with Texas across the board. It held that the World Court had no authority whatsoever to bind the U.S. justice system. At the same time, it struck down the president's order, concluding it was unconstitutional for the president to unilaterally surrender the sovereignty of the United States of America.

Today, the Court's decision in *Medellín* is especially important. At the time of this writing, the Obama administration is threatening to try to go to the United Nations to end run Congress and get a Security Council resolution adopting whatever Iran nuclear deal they are able to negotiate. The purpose, they have explained, is to try to transform that UN resolution into domestic law that would bind a subsequent president. But, as the Court held in *Medellín*, the UN has no authority whatsoever to bind the United States. President Obama's UN-Iran strategy cannot succeed, because *Medellín* made explicit that no president—Republican or Democrat—has the constitutional authority to subvert U.S. sovereignty.

The *Medellín* victory was profound for me because our argument had gotten to the heart of how I understood our government. We had been opposed by briefs from ninety foreign nations, the European Union, experts on the World Court, and the American Bar Association. Our opponent, José Medellín, was represented by one of the biggest law firms in the world. And yet none of these opponents were as formidable as our biggest adversary, the president of the United States, represented in court by one of the best appellate advocates in America.

But we had prevailed. Fifteen years after the rape and murder of two girls in Houston, the case of *Medellín v. Texas* ended with this simple declaration from the pen of the Chief Justice of the United States: "The judgment of the Texas Court of Criminal Appeals is affirmed. It is so ordered." On August 5, 2008, at 9:48 p.m., the state of Texas injected a dose of lethal chemicals into the bloodstream of José Ernesto Medellín.

Among the witnesses at his execution were the families of Jennifer Ertman and Elizabeth Pena.[30]

———

Unfortunately, *Medellín* was not the only case in which I found myself litigating in the Supreme Court in opposition to the Bush administration. Almost from the start, the administration had showed a willingness to oppose conservatives in high-stakes litigation. For example, in 2003 the U.S. solicitor general shocked observers by arguing for a "middle ground" on affirmative action, expressly accepting the permissibility of the government imposing race-based preferences. He agreed with conservatives that the University of Michigan's expansive affirmative action programs violated the Fourteenth Amendment's guarantee of "equal protection of the laws," but he also claimed that *some* government affirmative action programs are constitutional, even though the color of an applicant's skin is an explicit factor in deciding on admissions.

In 2008, the administration went wobbly again, and once again the Texas Solicitor General's Office was called to serve as a check and to defend conservative principles. The case was *District of Columbia v. Heller*. The issue was the constitutional right of Americans to keep and bear arms.

At the time, Washington, D.C., had the most draconian gun control laws in the country. The city prohibited any resident from owning a handgun, and it required that long guns be kept inoperable at all times, without exceptions. The ineffectiveness of this thirty-year policy was perhaps best evidenced by the city's murder rate, which in 2003 was worse than war-torn Baghdad's.

In a debate with the District of Columbia's attorney general on *The News Hour with Jim Lehrer*, I pointed out that residents in wealthy D.C. suburbs like Bethesda, Maryland, can purchase effective police protection through their tax dollars, and so it is easy for some of those residents not to value their Second Amendment rights. But for an African-American single mother who takes the subway in the middle of the night from her second job to her apartment next to a crack house in inner-city Anacostia, the right protected by the Second Amendment is far from abstract. With good reason, she may believe she needs a firearm to protect herself and her children. She is among the most vulnerable members of the American family, and when you take away the Second Amendment right of law-abiding citizens to "keep and bear arms," you take away the ability of Americans like her to defend themselves.

The D.C. attorney general was rather dismayed at that line of argument. He seemed to view it as unacceptable for a conservative Republican to defend the vulnerable. But gun rights represent only one of the many areas in which what I call "opportunity conservatism" empowers the least fortunate among us in a way that regulations and bureaucrats never have and never will. At every opportunity for public debate, it's our job as conservatives to focus on personal choice and empowerment.

Of course, protecting gun ownership isn't just a matter of good public policy. It is also required as a matter of constitutional law. So when the constitutionality of D.C.'s gun control laws arrived at the Supreme Court for argument in 2008, I was optimistic that Solicitor General Paul Clement would argue for a robust interpretation of the Second Amendment.

Sadly, the Bush administration did not allow him to do that. The Department of Justice refused to support Dick Anthony Heller, a federal law enforcement officer and D.C. resident who was challenging the city's prohibition of handguns. Instead the administration argued that "reasonable restrictions" are constitutional if they protect "important regulatory interests"—whatever that means. The District of Columbia's attorney general went even further, audaciously contending that the Second Amendment offers no protection whatsoever to individual gun owners, because according to the district, it protects only the "collective right" of militias.

I was dismayed with the Bush administration's attempt to water down the Second Amendment and incensed with D.C.'s attempt to write the Second Amendment entirely out of the Constitution. So was Bush's own vice president, Dick Cheney, who as president of the Senate signed on to a robustly conservative brief filed by 55 senators and 250 congressmen.

Texas took the lead among the states defending the Second Amendment. In the U.S. Court of Appeals for the D.C. Circuit, I presented oral argument in the companion case to *Heller*. And, before the Supreme Court, we wrote an amicus brief joined by thirty other states, in support of Heller's challenge to the district's laws.

Texas was willing—indeed, eager—to say that those laws violated the plain language of the Constitution. Unlike the District of Columbia, we did not believe the Second Amendment applies only to militias. And unlike the Bush administration, we did not believe that laws infringing Americans' right to "keep and bear arms" become constitutional whenever a federal judge finds them "reasonable." That's not what the Constitution says; instead, it says "the right of the people to keep and bear arms *shall not be infringed*." In my view, "shall not be infringed" means exactly that.*

In June 2008, the U.S. Supreme Court agreed with Texas. It explained that "the enshrinement of constitutional rights necessarily takes certain policy choices off the table." Among those unconstitutional policy choices was D.C.'s "absolute prohibition of handguns held and used for self-defense in the home." The merits of gun control may be debatable, "but what is not debatable is that it is not the role of this Court to pronounce the Second Amendment extinct."

The decision was 5–4, which meant that four justices had

---

*That doesn't mean that there can be never be any restrictions on gun ownership. For example, all of the amici states (and all nine justices) agreed that the long-standing prohibitions on felons owning firearms are consistent with the original understanding of the Second Amendment. But any restrictions must meet a strict standard, consistent with the Constitution.

agreed with the District of Columbia's radical position: not merely that some forms of "reasonable" gun control laws are permissible, but rather that nobody has *any rights whatsoever* under the Second Amendment. In other words, four justices would have held that the Second Amendment protects no individual right at all, that it establishes merely a "collective" right and is hence unenforceable by any American. Under that extreme interpretation, Congress could pass legislation making it a criminal offense for any American to own a firearm, and no man or woman in the land could challenge that law. And, ominously, we were just one vote away from the Supreme Court adopting that position, effectively erasing the Second Amendment from the Bill of Rights. If that doesn't highlight the importance of the next president's Supreme Court nominees, I don't know what does.

Once again, the state of Texas had defended conservative principles before the Supreme Court. And once again, a majority of the Court had vindicated that defense.

It was a sweet note on which to end my five and a half years as the Lone Star State's solicitor general.

———

When I initially took the job as solicitor general, Attorney General Abbott asked me to commit to staying at the post for at least two and a half years. I gladly agreed, and I intended to return to private practice after that period concluded. The reason that I stayed for more than twice the amount of time I had initially expected was simple: I loved the job.

Even the cases covered in this chapter do not fully capture how intellectually rewarding it was to serve as my home state's solicitor general. In addition to those cases, we litigated and won hundreds more, including successfully defending, as amicus before the U.S.

Supreme Court, the federal partial-birth abortion law, New Hampshire's parental notification law, and Indiana's voter ID law. And in the lower courts, we successfully defended Texas's law prohibiting taxpayer funding for institutions that provide abortion, the Texas marriage laws, and the Texas moment of silence legislation.

And, before the Texas Supreme Court, I argued and won unanimously a case upholding the Texas Sexually Violent Predator Civil Commitment law, which had been struck down by a state court of appeals. No cases were more horrifying than the numerous criminal cases I litigated involving rapists and pedophiles—I'm convinced there's a special circle of Hell for such coldhearted predators—and helping play a meaningful role preventing sexual assault and child molestation made my work in law enforcement especially rewarding.

All of these cases were handled with the assistance of an amazing cadre of professionals. The attorney general's office has roughly four thousand employees, and a budget over $500 million. When I arrived, I was put in charge of supervising every single appeal, criminal and civil, state and federal, throughout the agency. It was my first experience in a major executive position. I had led significant teams at the Justice Department and the Federal Trade Commission, but now I had supervisory responsibility for thousands of appeals being litigated by more than seven hundred attorneys. I worked hard to overhaul the reporting mechanisms, recruit new talent, and put in place senior attorneys to help manage the process. And, over five and a half years, we compiled a remarkable record of success.

I know of no other job like that of being solicitor general of Texas. Every day, you find yourself drawn into important litigation that is strategically difficult and fascinating—from congressional redistricting to the Second Amendment to the separation of powers. Fantastic cases land on your desk one after the other. And in every case, Attorney General Abbott gave me the mandate to fight for the conservative principles that make our Constitution enduring and our country free.

With his leadership and the work of an extraordinarily talented office of appellate advocates, that's exactly what I tried to do. And what I hoped to do in the future.

My tenure also brought into sharp relief the growing divide between Republicans in Washington and conservatives everywhere else in America. An increasing number of Americans, especially in the bourgeoning tea party movement, could not understand how the people we'd elected to office to articulate and defend conservative principles seemed instead to be drawn in, becoming more and more a part of the problem.

CHAPTER 7

Question #10: A Grassroots Campaign for the U.S. Senate

Pat Brown was elated. The powerful two-term Democratic governor had once been worried that George Christopher, an appealing Rockefeller Republican, friend of Soviet leader Nikita Khrushchev, and mayor of San Francisco, would prevail in the race for the Republican nomination to challenge Brown's reelection. Christopher, the Brown team believed, was the most dangerous kind of Republican candidate, a man backed by the establishment who didn't make any waves.

In June 1966 Republican voters thought differently about who was the best bet to beat the venerable Democrat. They selected as their nominee the candidate of Pat Brown's dreams—the former actor and corporate pitchman Ronald Wilson Reagan.

In Brown's view, Reagan, who'd made national headlines in 1964 by giving an impassioned televised speech for the party's failed presidential nominee, Barry Goldwater, was just the same sort of dangerous, delusional reactionary who had cost the GOP the White House in the ensuing Lyndon Johnson landslide. Reagan also, Brown believed, was a dimwit, a genial empty suit whose best-known film at the time was one in which he costarred with a monkey—*Bedtime for Bonzo.*

Reflecting the same mind-set, the legendary movie honcho Jack Warner, who had built countless movie star careers by casting them in major films, was said to be aghast at a Reagan bid. "No, no. Jimmy Stewart for governor. Ronald Reagan for best friend."[1] Even the candidate was in on the joke. A bright, thoughtful man who'd honed his free-market philosophy for decades, he reveled in the chance to be underestimated by his opponents. His greatest asset in that effort was his modesty and sense of humor.

Asked in 1965 what kind of governor he would be, Reagan replied, "I don't know. I've never played a governor." When a reporter asked for an autograph on a studio photo of Reagan and Bonzo, Reagan complied, writing above his signature, "I'm the one with the watch."

Many establishment Republicans in Washington also scoffed at Reagan's candidacy. New York Republican governor Nelson Rockefeller alleged Reagan was a political chameleon. House Minority Leader Jerry Ford tweaked Reagan's reputation for ideological stubbornness, "I don't think it's a great mark of character to put your feet in cement and then stand there." Richard Nixon, the former vice president, did not hide well his view that his fellow Californian was shallow.

What the mainstream politicos did not understand about Ronald Reagan was that he was running for a reason—not for money and not for fame (he already had both). He was running to change the country. Even more galling to the professional class that worked hard to polish the rough edges off their candidates for office, Reagan intended to be who he was.

He's been a proponent of the far right for years, Brown charged. Some Brown supporters retouched photographs to show Reagan looking like Hitler. Almost as if it were a dare, Brown declared, "I'll be very surprised if even he would deny that he is a conservative of the right-wing school."

The fifty-five-year-old Reagan had no intention of denying that. "You have a right to know—and I am obligated to tell you—where I stand and what I believe."

He was going to tell the truth about the state of the country and of California, which in the sixties was being torn apart by race riots, violent antiwar protests, the drug culture, and a general sense of gloom. He knew that Democrats outnumbered Republicans by 3 to 2 in the state, but he also understood that his basic arguments had a widespread appeal. That he could build a coalition not by hiding his real views, but by explaining them in terms people could understand.

As Reagan himself later put it, "People were tired of wasteful government programs and welfare chiselers; and they were angry about the constant spiral of taxes and government regulations, arrogant bureaucrats, and public officials who thought all of mankind's problems could be solved by throwing the taxpayers' dollars at them."

His campaign statements were pointed, memorable, and transformative. And he was a happy warrior, who used humor and self-deprecation to make his case. From the campaign trail, *Time* magazine reported, "Reagan can hold an audience entranced for 30 or 40 minutes while he plows through statistics, gags and homilies." He spoke to a wide range of Californians who'd had enough of big-government solutions. A sampling:

- "In California, government is larger in proportion to the population than in any other state and it is increasing twice as fast as the increase in population. Our tax burden, local and state, is $100 higher per capita than it is in the rest of the nation, and the local property tax is increasing twice as fast as our increase in personal income."

- "We see today a second generation, and even a third generation of citizens, growing up, marrying, having children, accepting public welfare for three generations as a way of life. The 11th century Hebrew physician and philosopher, Maimonides, said there are eight steps in helping the needy. The lowest of these is the handout; the highest is to teach them to help themselves."

- "We've had enough talk—disruptive talk—in America of left and right, dividing us down the center. There is really no such choice facing us. The only choice we have is up or down—up, to the ultimate in individual freedom consistent with law and order, or down, to the deadly dullness of totalitarianism."

- "Do we still have the courage and the capacity to dream? If so, I wish you'd join me in a dream. Join me in a dream of a California whose government isn't characterized by political hacks and cronies and relatives—an administration that doesn't make its decisions based on political expediency but on moral truth. Together, let us find men to match our mountains. We can have a government administered by men and women who are appointed on the basis of ability and dedication—not as a reward for political favors."

Slowly his efforts gave rise to a grassroots movement across the state. The Reagan registration campaign encouraged supporters to register six new voters each. The goal was a million new registrations, which as it happened was the exact margin of Reagan's landslide over Brown—in what one biographer later called a "stunning, out-of-nowhere victory." A large number of Democratic voters crossed over for this so-called right-wing extremist, including nearly one-third of labor union members. It was the beginning of a force that would transform political history: Reagan Democrats.

Pat Brown was shell-shocked by the result. A man once touted as a potential presidential contender lost in a landslide he never contemplated. But Reagan had seen it coming. Brown, Reagan recalled, was like many politicians who had become part of the establishment. He had "a kind of arrogance that takes place when an administration grows old and tired in government—an arrogance that makes someone have a contempt for the intelligence of the people. And I don't think he's been able to hide that." The Reagan Revolution had begun.

―――――

As my career back in Texas began to take root, Heidi and I faced a decision that confronts many families with two professional careers. For nearly two years, we had commuted cross-country, flying back-and-forth every weekend, but that was hard on a young marriage; so we decided together that our next chapter would be fully in Texas. But leaving D.C. wasn't easy for Heidi. She was flourishing in the Bush administration; she loved her job at the National Security Council, which enabled her to focus on many of the issues in the developing world that had so inspired her when her parents were missionaries in Africa. Heidi had always planned to combine a career in public service with substantial years in business. And this was a good time after Bush's first term for her to return to the private sector. But Heidi wasn't a native Texan, and it was initially hard for her to move to a new state, away from so many people she loved. The adjustment led to her facing a period of depression, which was really difficult for us both. I did my best to help Heidi through this time, we prayed together and she went to counseling and relied on the love and support of her family and close friends. In the end, it was a period that strengthened our marriage and was an important spiritual turning point for us both. God can use people who want to serve Him, regardless of location or circumstance, if we are open to His plan. Some months later, a friend invited Heidi to a Christian retreat—called the Road to Emmaus, named after the road described in Ephesians where Jesus appeared to the disciples after he had been crucified. He walked there with them and they did not even know it was Him for an extended period. That retreat helped Heidi turn the page and embrace our next chapter.

Like the hard charger she is, she built a thriving business with Goldman Sachs in Texas. Now Goldman is often a target for political criticism, and when they or any other big banks are coming to Washington seeking special favors, they deserve the criticism. But Heidi's job had nothing to do with Washington; instead, she was working with Texas families and foundations, helping them invest and save for the future. And she was really good at it; her clients trusted her. If you're handing over your life savings to someone to invest, what matters most is that you can trust them to look out for you. Heidi earned her clients' confidence, because she's the most trustworthy person I know. She also came to love the state as much as I did.

Within a few years, we started a family.

On April 14, 2008, Heidi and I welcomed our first child into the world. A precious—and now precocious—baby girl, Caroline

Camille Cruz captured my heart from the first moment I saw her. She shows no sign of letting it go. As she lay in the hospital bed, in labor, Heidi was typing furiously on her Blackberry, still tending to the needs of her clients. I admired her tenacious work ethic—it's one of the many qualities that made me fall in love with her—but this was too much. I gently pulled the Blackberry out of her hands. "It will be here later," I said. She had more important things to do.

To be fair, when it came to leaving work at the hospital steps, I wasn't completely innocent. During much of the time we were there, I was studying cases for an oral argument before the U.S. Supreme Court scheduled for two days later. I was appearing in support of a Louisiana law that allowed capital punishment for the very worst child rapists. It was a horrible case, where a three-hundred-pound man had brutalized his seven-year-old stepdaughter. So just hours after Caroline was born, I said a prayer of thanksgiving, kissed my beautiful wife and baby daughter, rushed to the airport, and flew to Washington to argue the case.

As with many young families, the birth of our first child led us to reexamine our financial circumstances and our long-term career goals. Heidi knew that I envisioned a future in public service, and this seemed like an opportune time to try to establish some financial security. I decided that it was time for me to move on from the solicitor general's office to return to the private sector.

I interviewed with a number of Texas firms and ended up joining the Houston office of Morgan Lewis & Bockius, a large international firm formed in Philadelphia in 1873. Today the firm has nearly two thousand lawyers and well over $1 billion in annual revenue; it is one of the largest law firms in the world. Morgan Lewis had historically been an apolitical firm, though one of my future partners was Fred Fielding, who had the distinction of serving as White House counsel for two presidents—Ronald Reagan and George W. Bush.*

I joined the firm as a partner and was given the responsibility of leading its U.S. Supreme Court and national appellate practice.

When I interviewed with various law firms, I was candid about my intention to run for elected office in Texas. The dissatisfaction of the American people with both political parties was increasingly clear in the later years of the Bush administration—a dissatisfaction that grew more acute with the financial collapse of 2008 and Washington's bailout of major Wall Street banks. As much as I respected President Bush, the bailouts were just wrong.

The Bush administration came to office promising to revitalize the Reagan Revolution—which had inspired me to get into politics in the first place. Instead, it took the Republican Party down the path of bigger government, excessive spending, and new entitlement programs that we couldn't afford.

By 2008, many of us were disenchanted. A Republican president should not add $5 trillion to the national debt. There was the controversy over the selection of Harriet Miers to the U.S. Supreme Court; the seemingly ineffective response to Hurricane Katrina; supporting the legendary "bridge to nowhere" in Alaska, which became an emblem of Washington waste; and an immigration plan that appeared to promise amnesty to illegal immigrants while failing to solve the fundamental problem. All of these things ran contrary to the kind of administration the Bush team had promised eight years earlier.

I could support Bush's "compassionate conservatism" because I

---

* In his office Fielding had two of the coolest pictures I had ever seen. One was of himself on Air Force One standing next to President Reagan, who held a bumper sticker reading, "My lawyer can beat your lawyer." The second was of himself with George W. Bush, on Air Force One, holding up the very same bumper sticker.

believed it reflected my belief that conservative policies are those that best help the aspirational goals of every American. Conservative policies—what I call "opportunity conservatism"—empower small businesses, encourage job creation, and help people lift themselves from poverty. I did not think that compassionate conservatism should become an excuse for government to intrude more and more into our everyday lives.

As the Bush administration drew to a close and the 2008 campaign was in full swing, it was not the moment for me to go into politics. But my partners at Morgan Lewis had no objection to my planning to do so in the future; in the meantime, I was charged with building a top-flight appellate litigation team that could endure at the firm even if I did get elected to office. I set about that task at once.

As solicitor general, I already had experience managing the appeals for more than seven hundred attorneys. Morgan Lewis had a tremendous global practice, representing over half of the Fortune 100 companies, but it had not historically had a strong Supreme Court practice, which is a hallmark of most top-tier firms. My job was to change that.

I began by recruiting some of the best talent I could find, bringing in skilled appellate litigators who had clerked for Justice O'Connor, Justice Scalia, and Chief Justice Rehnquist. And then we set out building the business, pitching new clients and growing the practice. Our focus was on "bet the company" cases, representing major corporations in cases that few observers thought were winnable.

I represented FedEx before the U.S. Court of Appeals for the D.C. Circuit, winning a major victory and overturning an adverse decision from the National Labor Relations Board. Other significant victories included successfully representing Kimberly-Clark, Dentsply, and AstraZeneca in high-stakes commercial litigation.

I also represented a French housewares company, SEB, in multimillion-dollar patent litigation before the U.S. Supreme Court against a Chinese company that had stolen their intellectual property (a patented deep fryer), reverse-engineered it, and sold thousands illegally in the United States. Arguing a case before the U.S. Supreme Court in the middle of a campaign for Senate was not easy—the time it took for me to prepare drove my campaign team nuts—but the results were worth it. When the Court took the appeal, the widespread assessment from observers was that SEB was likely to lose, and yet we ended up winning the case 8–1.

Two of the most significant cases I litigated were both pro bono, which meant the firm and I handled them for free. I was deeply honored to represent more than three million veterans defending the constitutionality of the Mojave Desert Veterans Memorial, a lone white Latin cross that had been erected over seventy years earlier to honor our soldiers who gave their lives in World War I. The American Civil Liberties Union had sued, and they had won—the district court and the court of appeals had both ordered that this long-standing veterans memorial be torn down, because they believed the image of a cross could not be allowed on federal land. On appeal, I represented the American Legion, the Veterans of Foreign Wars, and several other veterans organizations before the U.S. Supreme Court, and we won 5–4.

That victory was yet another opportunity to defend religious liberty, a deep passion of mine, and, today that monument to our brave soldiers still stands.

I also was proud to represent John Thompson, a Louisiana man who had been wrongfully convicted of murder and sentenced to death. Two of my partners had represented Thompson pro bono for decades, and they had uncovered DNA evidence that proved his innocence. Tragically, the Louisiana district attorney's office had deliberately suppressed the DNA evidence, and Thompson spent eighteen years of his life imprisoned for a crime he did not commit. He was released, and he subsequently sued the DA's office for their wrongful conduct. A jury awarded him $14 million, and I helped represent Thompson on appeal.

Some caricatures suggest that a conservative would be reluctant to represent a convicted murderer. That may be true, if the client is clearly guilty. Although every defendant deserves a lawyer, I've handled too many horrible criminal cases to have any interest in representing violent criminals. But John Thompson was innocent. And critical to supporting the death penalty is ensuring that we vigorously protect the innocent. DNA has enabled many guilty persons to be convicted, and it has proven the innocence of many others.

Before the court of appeals, we won with an equally divided court, 8–8. One of my partners then ably argued the case before the U.S. Supreme Court, and, unfortunately, we lost, 5–4. Thompson deserved restitution for the injustice that was done to him, and it saddened me to see him denied. But, I must say, it was inspirational to spend some time with Thompson, to see the grace and peace he manifested despite having nearly two decades of his life stolen from him; upon release, he started a nonprofit to help others exonerated from wrongful convictions.

Each of these cases in private practice had a significant impact. And, over five years at Morgan Lewis, our appellate practice grew to more than thirty attorneys nationwide, became a significant revenue generator for the firm, and racked up a remarkable record of wins, so much so that the *National Law Journal* in 2010 included Morgan Lewis for the first time on its "Appellate Hot List," their ranking of the top twenty law firms "that represent the best in the practice of appellate law."

———

In 2009, Texas politicos were undergoing their unique version of musical chairs. U.S. senator Kay Bailey Hutchison had stated that she intended to resign her Senate seat and run for governor against the Republican incumbent, Rick Perry. David Dewhurst, the lieutenant governor, in turn planned to run for Hutchison's open Senate seat, which left Attorney General Greg Abbott free to run for Dewhurst's seat as lieutenant governor. I decided to campaign for attorney general, to replace my former boss, and I declared my candidacy for the post in January 2009.

At the time, there were four serious Republicans looking at the attorney general's race, five if you included myself as a serious candidate—which most Texas observers did not. I could understand why. I had never been elected to any political office. I'd never even run a campaign. I had zero statewide name recognition. Indeed, as I often joked, the last elected office I had held was on student council.

My likely opponents included two state supreme court justices (both already elected statewide), a member of Congress personally worth hundreds of millions of dollars, and a state representative who represented and was beloved by the biggest donors in Dallas.

If I had any hope of winning, we had to do something that had never been done before—build a massive grassroots army to overcome the money and vested interests across the state.

My first task was assembling a team. The very first person who joined me was my political consultant, a young man named Jason Johnson. Jason had some of the characteristics of a Republican James Carville—though I doubt either one would appreciate the comparison.

Jason came from modest circumstances in rural East Texas. When I made reference to the fact that he'd grown up in a double-wide trailer, he feigned offense—because his trailer wasn't double-wide. Those, he said, belonged to the rich people in town.

I first met Jason when he was Greg Abbott's top political strategist. At the time, he had run nearly 70 campaigns in Texas and had won all but two of them. He has a deep, analytical sense for politics and an instinctive sense about people—two priceless qualities in a political strategist. The *New York Times* once described him as "a scrappy East Texan" with "a touch of mad genius." That led me to give him the nickname "Touch of Genius."

Having forged a well-deserved reputation as a top political consultant, Jason could have worked for any of the candidates considering the race. Indeed, conventional wisdom was that the right thing for him to do was to sign on with one of the candidates with un-limited financial resources, who could compensate him richly for his services. Jason decided to come with me instead. We knew each other well, and he cared a good deal that the next attorney general, like Greg Abbott, would be a principled conservative who would stand and fight for the causes we believed in. My hiring of Jason was the first signal to Texas politicos that I was serious about the race.

My second hire was John Drogin, a former reporter at the Vat-ican who had served as press secretary for the then-junior senator from Texas, John Cornyn, and thus knew many reporters in the state well. Drogin and I were introduced through mutual friends and met at a coffee shop where we spent several hours getting to know each other. I liked him immediately—so much in fact that at the end of our interview, I extended an offer to him to become my campaign manager.

His response revealed a lot about his character, and convinced me I had made the right choice. Drogin, who at the time was jobless, didn't just jump at any offer to manage a statewide campaign. In-stead he said, "I need to ask you two questions before I can accept."

The first question was "Are you a Christian?" The second was "Are you pro-life?"

He had had his heart broken before by politicians who said one thing and then did another once elected. "I want to look you in the eyes when you answer both of these questions," he said. The earnestness with which he said this was endearing, and thankfully I was able to answer both questions in the affirmative.

With Johnson and Drogin on board, we built a small, incredibly lean campaign of young people committed to real change and to working harder than anybody else. That's how you build a grassroots campaign—with commitment and shoe leather. Over the next year, we attended hundreds of small events across the state of Texas. I sat with people, listened to them, and shared my beliefs and vision. So many of the Texans I encountered, from both parties, had grown tired of career politicians who said what voters wanted to hear, promised them the moon, and then disappointed them, because the politicians weren't truly committed to real change. I was determined to show them that I was not one of those people.

The momentum we built was slow but steady. With each person to come on board our campaign, it became easier to recruit the next person and the next. And we did on occasion receive boosts from unexpected corners.

For example, Heidi and I became friends with George P. Bush, the grandson of George H. W. Bush, son of Jeb, and nephew of George W. I liked and admired "P," as friends called him. Growing up in a famous family that had seen such tremendous success could cause anyone to be full of himself, but George P. was always well grounded and possessed a remarkable sense of humility. And, like me, he married up: his wife, Mandi, is beautiful, brilliant, down-to-earth, and charming.

During the attorney general campaign, George P. generously

asked me if I would like to visit with his grandfather up in Kenne-
bunkport, Maine, where the Bushes spent their summers. Of course
I would.

On my way up from Texas, I spent a lot of time thinking about
what I wanted to say to the former president. I did not know him at
all. So I decided I wasn't going to ask him for anything in my race.
Instead I was just going to sit with him and solicit his advice—a
novice, aspiring candidate looking for some wisdom from a longtime Texas politician and American statesman.

Meeting with him in his office, I was surprised by how much he
already knew about my campaign and me. George P. had laid a lot of
positive groundwork. After we talked for about twenty minutes, he
asked, "So what time is your flight? Can you come out on the boat
with Bar and me?"

I responded that my flight back to Texas was in two hours, but
that flights could be changed quite easily. "Mr. President, I'll stay as
long as you'll have me."

He smiled and then scrutinized my attire—a suit and tie that I
felt appropriate to the occasion. "Ted, that's not going to do at all,"
he said. The former president drove me in a golf cart to his residence, where he fished out some jeans, a shirt, and a belt, the buckle
of which read, "President of the United States." It was surreal to
be wearing his clothes.

Then we met Mrs. Bush and went out on his boat. George Herbert Walker Bush was then eighty-four years old, but he drove that
boat like he was still an eighteen-year-old Navy fighter pilot. He had
the boat at full throttle, crashing into the waves, and spraying mist
into our faces with unrestrained glee. Barbara sat in the front of the
boat, with water pelting her face, and smiling as if nothing could
make her happier.

We made our way down the Maine coast to a small restaurant

where we dined on lobster rolls. On the way, we discussed the Green Revolution, which was just then playing out in Iran; we both were dismayed that President Obama was allowing to slip away this historic opportunity to support the democratic opposition to the virulently anti-American, theocratic rule of the Iranian mullahs.

At the end of our meal, we came back to the compound, where the former president had one more surprise in store. He reached into his pocket and pulled out a check from "George and Barbara Bush" in the amount of one thousand dollars and made out to the Ted Cruz for Attorney General campaign.

There have been few times in my life when I've been speechless— often to the great frustration of my critics. This, however, was one of those times. Eventually I stammered out a thank-you. I had to resist every urge to give the graying forty-first president a hug. I shook his hand, retrieved my rental car, and headed back to the airport, giddy that I had unexpectedly received the support of the patriarch of Texas's most influential political family. I called Heidi and told her in wonderment, "I just had the most magical afternoon."

After I flew back to Houston, my campaign followed up with the former president's office to ask if he might be willing to give a formal endorsement to our campaign along with his contribution. His aides asked us to prepare a draft statement, which we did.

A couple of days later, I received a phone call from Karl Rove. Karl and I had always been friendly. When I was on the policy staff of the George W. Bush campaign, before Karl became a "master of the universe," as he's depicted in the media today, he was a charming, gregarious guy who would put members of the policy shop in headlocks and playfully dubbed us "the propeller heads."

A couple of years earlier, I had sat down for a four-hour breakfast with Karl. I asked his advice on eventually running for office— whether I should stay on longer as solicitor general or go to private practice. Rove, an enthusiastic kibitzer on politics, generously shared his insights. He advised that I should stay on the job as solicitor general, keep building my record, and find opportunities to systematically build political support for a future run. It was good advice, and I tried hard to follow it.

Karl had found out about my meeting with George H. W. Bush and called me on the phone. He was irate, demanding, "What in the hell do you think you are doing?!"

It turned out Rove was in the process of helping raise money for the George W. Bush presidential library in Dallas. Texas donors were giving the Bushes tens of millions, including major donors who were supporting the Dallas state rep who wanted to run for attorney general. Those donors were now berating Karl.

"Well, Karl," I responded, "I was just doing what you suggested when we met. Going out and getting support."

"Yeah, well I didn't think you were going to get support from 41," he snapped. He suggested that the elder Bush was too old to have good judgment anymore. I was offended by that characterization, and knew from my visit with 41 that it wasn't remotely true.

As Karl continued to yell at me, I responded calmly, "Look, I got my start in politics working for Bush 43 and for you. . . . What would you like me to do?"

"Return the check," he said.

"Well, I can't do that. We already deposited it." I pointed out that under Texas's election law, we had to list the contribution on our ethics disclosure report.

He paused for a few seconds. "All right, fine. Then I want you to do nothing whatsoever to draw attention to it."

And then he pulled out the hammer. He implied that if I made any news about Bush 41's support, then Bush 43 would endorse my opponent and come out publicly for him—a threat that was fairly striking given that I had devoted four years of my life to working as hard as I could helping to elect Bush and serving in his administration. I always wondered whether Karl had the authority to make these threats on behalf of the former president—he certainly acted like he did. In any event, the last thing I wanted to do in running a fledgling campaign in Texas was to get on the wrong side of Rove and the second President Bush.

"Fine," I replied. "We'll do nothing to draw attention to it."

When I hung up the phone, I turned to Heidi, who'd been listening to the whole conversation. She was trembling, and visibly angry. At Karl. And at me, for caving in.

"You know what?" she said. "This is what's screwed up about the Republican Party. Why the hell should the Republican nominee for attorney general in Texas depend not on their qualifications, but on who the donors are to the Bush presidential library?"

A couple of hours after my conversation with Rove, we received an email from Bush 41's office. They had approved the draft endorsement we had sent, an unbelievably effusive statement from George and Barbara Bush calling me "the future of the Republican Party." I was grateful yet again. It was difficult to imagine that Bush 41 was unaware of the consternation that his endorsement would cause Rove.

In any event, I called Jason and told him to take our draft press release announcing the endorsement and throw it in the trash. We then informed Bush 41's office that I was immensely thankful for the support, that it meant so much to Heidi and to me, but we weren't going to release a statement. We didn't want to anger Karl or 43. The former president's office said he understood.

Our campaign's next hurdle was financial. We had to show that we could raise enough money to communicate with 26 million Texans. The only way to do that was to be able to afford statewide television ads. We set a goal for our first financial report to raise $1 million.

I spent the first six months of 2009 traveling the state, meeting with donors and asking for support. I didn't attack the other well-known candidates considering entering the attorney general's race; instead I sung their praises. They were friends whom I respected. But I also took the time to lay out why I believed I would do a better job.

When we met and exceeded our goal of $1 million, it sent shock waves across Texas. A political nobody had managed to pull what was widely considered a huge number.

We also launched a statewide leadership team that consisted of virtually every movement conservative leader in the state—people who had led grassroots groups. When I was solicitor general, we had fought side by side on many high-profile cases, such as defending the Ten Commandments and protecting U.S. sovereignty from the UN. Over time, these activists had come to know me and the sincerity of my beliefs.

As it happened, 2009 coincided with the first tea party rallies in Texas. The tea party was a true phenomenon of people fed up with politicians from both parties who'd raised our taxes, spent us into debt, and refused to listen to the voters who'd elected them. My father and I both spoke at some of the first rallies.

In Dallas, my dad told the crowd of more than ten thousand people that when he was a young man, he had seen a young, charismatic leader come to power promising hope and change, promising to redistribute the wealth. That man, of course, was Fidel Castro.

I posted a portion of my dad's speech on Facebook, and one left-leaning journalist wrote a blog post saying it was ridiculous for anyone to compare Obama to Castro. Lefty commentators went nuts. As I was reading the comments late one night, I decided to comment myself on the blog. I signed in and made the following observation:

"To all the commentators hyperventilating over my Father's speech, let me point something out—in his entire speech, he never once mentioned the words 'Barack Obama.' He simply described what Fidel Castro did in Cuba. Now . . . what does it say about you, that when you hear what Castro did, you immediately think it must be Obama?"

The unity of key conservative leaders in the state behind our campaign sent another shock wave through Texas politicos and journalists. And, over time, those developments began to suck the oxygen out of the race. As we'd hoped, one after another of the potential players looking at the race announced that they were not going to run. It now was clear that, in defiance of the odds and the professional political prognosticators, we were in an excellent position to win the attorney general's race.

That's when the floor fell out of our campaign.

In December 2009, Senator Hutchison announced that she wasn't going to resign her Senate seat after all. At that point, it was already becoming clear that Hutchison was likely to lose her bid to unseat Governor Perry. Her decision to stay in Washington meant that the game of musical chairs wouldn't happen. Everyone was stuck. Dewhurst could not run for Senate, meaning that Abbott could not run for lieutenant governor. When Abbott instead announced his campaign for reelection, I immediately ended my campaign and offered Abbott my enthusiastic support.

It was frustrating to have worked tirelessly for a year to build a formidable grassroots campaign, only to have the race disappear. But I hoped that our work would not be in vain. The following spring I sat down with Jason and John and spent several hours talking about the possibility that I might run instead for the U.S. Senate in 2012.

After her landslide loss for governor, Hutchison announced that she would serve out her Senate term, set to expire in 2012, but would not run for reelection. And the musical chairs began again.

On one level the idea that we might launch a campaign for the U.S. Senate was far crazier than my plan to run for attorney general. In fact it was audaciousness bordering on insanity. That wouldn't be the last time people would say that about me.

The reason was that my likely opponent was David Dewhurst, the sitting lieutenant governor. In some states that might not be so bad, but in Texas, the lieutenant governor has long been called the most powerful elected official in the state. The reasons stem back to post–Civil War days, when carpetbagger governors ruled with an iron fist. After Reconstruction, Texas responded by adopting the Constitution of 1876, a document that deliberately created a very weak governorship in order to prevent any future gubernatorial dictators. That left the lieutenant governor with enormous influence.

The lieutenant governor presides over the Texas Senate, and wields power over that body not unlike the Speaker of the U.S. House. He can dictate which committees consider which bills, which bills in turn are considered on the floor of the senate, and (often) which bills pass and which fail. If my opponent were the lieutenant governor, there was no doubt that everybody—absolutely everybody—who had business before the Legislature would be against me. That meant every major donor, every major trade group, every corporation, and every lobbyist in the state. And just about every elected official. They had to support my opponent, or else risk a crippling retribution.

Moreover, the sitting lieutenant governor was no ordinary politician. Dewhurst, who had been in office for a decade, was immensely wealthy. To his credit, he had made his money himself, through a successful career in business. And he had demonstrated in past elections a willingness to spend vast sums from his $200 million fortune in order to guarantee a win. If he ran (and he was almost certain to run), we could expect our opponents to have resources far greater than whatever we could raise.

Dewhurst also had universal name recognition: 96 percent. Virtually every Republican primary voter knew his name and had voted for him before. In a state the size of Texas, name identification is a big, big deal, because building it is incredibly expensive.

There was also a sense that he was entitled to the seat. He was "next in line," which in Republican primaries often matters a great deal.

Though I had gained some respect within Texas political circles from my attorney general bid, outside of those circles I was still unknown. Indeed, the first poll we conducted had us within the margin of error—not of David Dewhurst, but of zero. The name "Ted Cruz" registered the support of 2 percent of voters. The margin of error on the poll was 3 percent, which led Heidi to helpfully point out that, technically, my support might actually have been negative 1.

Early on in the Senate campaign, I sat down with Karl Rove once again, who'd cooled down considerably since yelling at me for courting George H. W. Bush.

Rove was never one to mince words. "Ted, you're a fool," he said. "You have no prayer of winning this race. Cannot be done." Referring to the several other conservative candidates who were planning to challenge Dewhurst, he said, "You'll be outraised by all of them." At that point I laughed out loud. I said, "Karl, you may be right that we can't beat Dewhurst, but I promise you you're not right in that. If you believe that they're going to outraise us, you've been out of Texas too long."

In fact, it had been a long time since Karl had been seriously involved in Lone Star State politics. He had been casually advising Hutchison in her race against Governor Perry, which she'd lost by 20 points. But the bulk of his time was spent helping lead the super PAC Crossroads, which many saw as trying to get moderate Republicans elected across the country.

"It doesn't matter," he said. "You're a fool because you've already won the AG race." His point was that the attorney general position would be mine for the taking, when Greg Abbott stepped down from the post to become lieutenant governor after Dewhurst was elected to the Senate. "You're blowing a sure thing and taking on an impossible race."

I told him, "You know what, Karl? You've been incredibly successful politically. You could be an elder statesman and make a real difference for the party. You know perfectly well that what the party needs in Texas and nationally. You could actually stand up, Karl, and do the right thing."

He just laughed and said, "Doesn't matter, can't be done. You don't have a prayer."

———

For a couple of decades, I've gone virtually every year to the Federalist Society's National Lawyers Convention in Washington, D.C., a gathering of conservative and libertarian attorneys, judges, and law professors. I have many dear old friends there. In November 2010, I made a new one.

At that particular lawyers convention, a newly elected senator named Mike Lee was in attendance. A Utah conservative who successfully challenged his state's incumbent Republican senator, Mike is the son of Rex Lee, a Reagan-era solicitor general who is widely considered one of the finest U.S. Supreme Court advocates in history. Mike followed in his dad's footsteps by excelling as a lawyer in a career that included a clerkship for Justice Sam Alito at the U.S. Supreme Court.

I ended up spending about three hours visiting with Mike. We went together from the convention in the Mayflower Hotel back to the Capitol. He was still down in the basement with temporary offices, and we walked the Capitol hallways talking for hours about legal issues, constitutional issues, and the challenges facing our country. We found ourselves agreeing on almost every topic we could come up with. I remember being struck that at some point he mentioned a recent study that had shown the massive value of the land owned by the federal government.

Only 13 percent of that land is federal parks. Most of the rest is in the American West, where states were forced to hand over a significant portion of their land as a condition of joining the Union. In Nevada, for example the federal government owns nearly *90 percent* of the land. This makes no sense. Parks are wonderful—and we need to preserve and improve them—but there is no reason for the federal government to own huge portions of any state. Nationwide, much of this land sits unused, while small parts of it are leased out for grazing or private use. Mike pointed out to me that the value of all that federal land was roughly $14 trillion.

At the time, the national debt also happened to be $14 trillion. That suggested to us an obvious and rather elegant solution for eliminating the debt and moving as much land as possible—other than national parks—into private hands.

Mike and I immediately bonded. A couple of months later, after I launched my Senate campaign, we spent another couple of hours talking in D.C. He said, "Well, is there any way that I can help you?"

I replied, "Sure, you could endorse me."

He said, "Done. I'm happy to." He looked at his staff, who were standing alonside, and asked, "I can do that. Right?"

After they nervously assured him he could, Mike said, "What else can I do?" I asked for his help reaching out to the other key conservative leaders in the Senate. He said, "Ted, I will move Heaven and Earth to get them to back you."

Over the next several months of 2011, Senators Rand Paul, Pat Toomey, Tom Coburn, and Jim DeMint all endorsed me. Of those five, the most impactful was Jim DeMint's, and the reason is simple: The senator from South Carolina had been, for a number of years, a lonely conservative voice in Washington, D.C., who spoke for those outside the Beltway. He had found in fight after fight that he had virtually no support in the Senate.

And so, in 2010, Senator DeMint decided to do something radical for a sitting senator. Something extraordinary. He decided that he was going to get involved in Republican primaries, because the only way to get the Senate Republicans to stand for conservative principles was to elect a different kind of candidate. He formed a group called the Senate Conservatives Fund, and Senator DeMint went on to play a critical role in helping elect Paul, Toomey, Lee, Ron Johnson, and Marco Rubio.

Every one of those senators might well have lost without the boost given by DeMint, who not only endorsed them but also began raising vast sums of money to help underdog conservative candidates defeat the choice of the party bosses in their primaries. The party bosses' choice can always be counted on not to rock the boat, and Senator DeMint understood that if we're going to change our ship's course before it careens over the waterfall, we need a lot of boat rocking to turn it around.

I first met with Senator DeMint in December 2010, before I had announced. DeMint had already endorsed Michael Williams for the Texas Senate race when everyone thought Kay Bailey Hutchison was going to step down—and when I was still running for attorney general.

Williams is a charismatic African-American conservative, a powerful speaker, and a good man, and he had already won statewide election to the Texas Railroad Commission. But when I sat down with Senator DeMint, I tried very hard to make the case for my candidacy. Number one, I thought my record of standing and fighting for conservative principles and winning was much stronger than that of my opponents. Number two, in order to beat David Dewhurst, you had to be able to raise enough money to run an effective statewide campaign, and I didn't believe the other candidates in the race could do so.

Senator DeMint listened and appeared interested in what I had to say, but he remained unconvinced. He told me, "Okay, I like what I'm hearing. Now show me. Go do it. Don't tell me you're going to raise money. Don't tell me you're going to build grassroots support. Don't tell me you think conservatives will unite behind the campaign. Go and do it, and then come back." I had the same reception from other major conservative groups, including FreedomWorks and the Club for Growth.

I was disappointed. I would have loved for Senator DeMint to say, "I'm supporting you right now." But what he was saying was exactly right. And today, when Senate candidates ask for my support, I say the same thing—show me. Build grassroots support. Raise money. Gain momentum. Put yourself in a position to win. And, then, if I can play a positive role and help push you over the edge, I'll do it gladly.

The gains made by conservatives in the 2010 midterm elections gave me confidence that a grassroots campaign for the Senate might work. I knew how tired people were of the usual politicians promising the usual things. There was a genuine sense of cynicism, even downright despair, about what was going on in Washington, D.C., now under a Democrat administration. I sensed an opening.

In our first benchmark poll, we asked a series of questions to assess where I stood compared to Dewhurst. One of those questions would become famous internally in our campaign: Question 10. It asked voters if they would be more or less likely to support me if they knew that "Ted Cruz understands that politicians *from both parties* have let us down. Cruz is a proven conservative we can trust to provide new leadership in the Senate to reduce the size of government and defend the Constitution." Among Republicans, those two simple sentences polled north of 80 percent. At the same time, they garnered a majority of Independents, and even 20 percent of Democrats. They became the centerpiece of our campaign.

As with the attorney general's race, a big question was money. Contrary to many pundits' misconceptions about the importance of money in politics, a candidate does not have to raise the most money to win—and we could never outraise a self-funder like Dewhurst. But we did need to raise enough to be heard. For a statewide U.S. Senate race, we calculated that we'd need a minimum of $5 million, and would ideally be in the ballpark of $10 million.

In that effort, we confronted a problem—the federal government's campaign finance laws. A great many people in Washington, backed by the media, proclaim the need to control the amount of money being spent on political campaigns. Campaign finance laws are the Holy Grail of so-called good-government types who want to do *something* to fix the problem. As is often the case in Washington, their solution makes things worse.

In Texas state government races, there are no limits on individual donations under state campaign finance laws. This had made it a lot easier for an unknown like me in the attorney general's race to raise money from a committed group of donors, compete, and potentially win against entrenched incumbents.

The former Reagan campaign consultant Ed Rollins tells a story of a group of eight California businessmen wanting to support Ronald Reagan's gubernatorial campaign. They sat around a table and asked how much the novice candidate and former B-movie actor might need to win the race. The businessmen were told he'd need at least $2 million. So all eight promptly wrote checks for $300,000 each, giving the Reagan campaign the infusion of cash it needed to compete. Those eight men—not household names, just California small business owners—changed the course of history.

By contrast, federal campaign finance laws make such an effort impossible in a race for the U.S. Senate. They impose strict limits on the amount any individual can contribute, in effect rewarding candidates with deep pockets who can self-finance (since there are no limits on what you can donate to yourself) or those who are already well-known across the state. Written by political incumbents, these rules function as incumbent protection laws designed to combat what they see as a great evil—that some outsider could raise enough money to defeat them.

As a result (and by design), it is practically impossible for someone who is not an incumbent politician—without an existing, massive fund-raising apparatus—to raise enough money in small increments to run statewide in a large state like Texas. To win, we had to fundamentally change the rules, and bring thousands of new players to the campaign.

As we were assessing the Senate race, I spent six months systematically sitting down with a couple of hundred potential donors and friends—over breakfasts, lunches, and coffees. I started out with the obvious—I was planning to run for the U.S. Senate against a man with universal name recognition and limitless funds. "Am I nuts?" I'd ask.

Almost all of them chuckled and said, "Yes." But in the end many added, "If you do it, I'm with you."

The strategy at the heart of the campaign was empowering people—to make ordinary citizens the central part of our crusade.

We decided to emulate an unlikely example: President Barack Obama. Thus we made the conscious decision to explicitly model our campaign on his 2008 primary campaign against Hillary Clinton. Back in 2008, Hillary Clinton was the most formidable primary candidate in modern history who was not an incumbent president. As with Dewhurst, the conventional wisdom was that she was unbeatable. But Obama ran a scrappy, grassroots, guerrilla campaign—phenomenal in the annals of politics—and beat her. He was the David to her Goliath.

We resolved to do the same. Indeed, I bought copies of Obama campaign strategist David Plouffe's book, *The Audacity to Win*, and gave it to our senior team. "We are going to shamelessly steal from their playbook," I told our team. Our goal was to build a grassroots army and make sure it was *their* campaign. With them, we would go to Washington to turn things around. And unlike Barack Obama, I knew I would keep my promise.

Through the years, Obama's use of the campaign slogan "hope and change" has won justifiable mockery. That's because it was only that: a slogan. But I too believed in hope. Our campaign was based on it. I knew that so many Texans, like the rest of our country, really believed in the Promise of America, really worried about its future, and were certain that we could find a way to restore America to the world's greatest, most respected, nation again. I also believe in change—but not merely a change of parties or a change on the surface.

Washington is broken, and our country is in crisis. To actually solve the massive fiscal and economic challenges facing America, we have to empower citizens to bring fundamental changes to the way Washington operates. I knew that David Dewhurst was a nice man, an honorable man. But he was not a person who would go to Washington and shatter china, challenge convention, and be willing to stand up to members of both parties—the kinds of actions that would be needed if we were really to change an entrenched system.

On January 19, 2011, we announced our campaign for the U.S. Senate in a rather unorthodox way: on a call with dozens of bloggers, rather than at a press conference with the usual political reporters.

The reason for our decision was that the mainstream media has a herd mentality. They all shared the conventional wisdom that Dewhurst would win in a walk, and they were sure to pleasantly ignore our campaign as long as they could.

But in an era of twenty-four-hour news and the Internet, the mainstream media is a sagging dinosaur. The world that we knew growing up—where three news networks and the daily paper decided what the news was—was over. In the age of new media, one blogger can be heard around the world. The representatives of new media would be the ones who would help us frame our campaign— as a battle between a traditional career politician and a proven conservative fighter.

If that became the message of the campaign, the race would be over. But hoping to frame a campaign is not the same as knowing how to frame a campaign. Voters are more sophisticated than pundits and politicians believe, and grassroots voters are especially savvy. For good reason, they wanted proof—repeated demonstrations—that I wasn't going to be a part of the establishment crowd.

One of the ways I tried to demonstrate I was different was by showing that I knew it takes more than just a good voting record if you're going to fight for conservative causes—it takes accountability. Thousands of times, I told voters across the state, "If I go to Washington, and I just vote right a hundred percent of the time, I will consider myself an abject failure." That's because we're in a time of crisis, and we need senators who will stand up and lead. "If I am not standing on the front lines with arrows sticking out of my torso, I won't be doing my job." (I didn't quite realize during the campaign just *how many* scars I'd acquire in the next couple of years.)

I told Texans over and over again, "I want you to hold me accountable. Hold me accountable for these commitments. When I come back in front of you in six months or a year, if I haven't done exactly what I said, if I'm not leading the fight to get back to the free-market principles and the constitutional liberties this country was built on, I want you to stand up and look me in the eyes and say, 'Ted, why did you break your word and why did you lie to me?'"

Throughout the campaign, we focused on Question 10, the one that took both parties to task for what was going on in Washington. Our opponents did not understand this strategy, which led to one of the more amusing moments of the campaign.

At a big rally in the blazing heat, on the steps of the state capitol, with Ron Paul, Rand Paul, and thousands of young people in attendance, I said something very similar to the message of Question 10: that career politicians in both parties are responsible for the mess we're in, and that we need the people to help get us out of it. We needed leaders who stood for Texas and who didn't just listen to Washington political bosses. Those assertions drew passionate cheers, just as they always did.

The Dewhurst campaign pounced on the comment, thinking I'd made a gaffe. They blasted out my statement to everyone. Their spin: "Ted Cruz is blaming Republicans for the problems in Washington. Can you believe that?"

It was a telling moment—reflecting the widening gulf between establishment Republicans and the grassroots. Their side viewed criticism of the GOP and its leaders as heresy. Our side saw it as a statement of the obvious.

What the Dewhurst side didn't understand, because they weren't talking to the same people, was that when I had said something like that to a meeting of the Republican Party in Houston, or to a Republican women's club in any county in the state, the response was a standing ovation. Everyone was sick and tired of Republican politicians saying one thing and doing another thing once they went to Washington. Of course, the only people who didn't know that were the establishment Republican politicians. The voters instinctively knew that big-government Republicans contributed to the problem just as much as Democrats, and in some ways even more, because the establishment Republicans lied to us over and over again by mouthing conservative platitudes that they didn't really believe or would never fight for.

In the early days of the campaign, I was often a lone traveler, with no driver and no scheduler. I climbed into rental cars and headed to dinners or events in small venues, meeting with a handful of local leaders a few dozen at a time. People like that wanted to get to know the candidates personally to find out what made them tick.

One of the points that I made often on the campaign trail was about my campaign website. Just about every political candidate for any office has an "issues page" on their campaign website. In Texas, that page was pretty predictable. If you took five Republican statewide candidates and printed out their respective issues pages, on just about every issue they would say the exact thing, almost word for word.

So I often pointed out to the grassroots activists that my website, TedCruz.org, did not have the customary issues page. In its place we put a page called "A Proven Record."

Our website explained, "Far too many candidates say one thing and do another. Every one of us should ask any candidate who stands before us, 'you say you believe these principles, *show me*. When have you stood up for them, when have you bled for them, and what have you accomplished.'"

Take, for example, the Second Amendment to the U.S. Constitution—the "right to bear arms." Every Republican candidate in Texas, and just about every Democrat, would say on his or her issues page, "I support the Second Amendment." In Texas, unless you are clinically insane, that's the right answer for a candidate.

But nowhere on my website did I say that I supported the Second Amendment. I said nothing about what was hidden in the deep recesses of my heart. Instead our website described how, as the solicitor general of Texas, I had led a coalition of thirty-one states before the U.S. Supreme Court defending the Second Amendment right to keep and bear arms, and we had won a landmark decision upholding that constitutional right. And we then linked to my being awarded the NRA's 2010 Carter-Knight Freedom Fund Award, given each year to one of the leading defenders of the Second Amendment nationally.

That was true on issue after issue after issue. I often quoted my former boss at the Department of Justice, John Ashcroft, who used to say, "If I'm ever accused of being a Christian, I'd like for there to be enough evidence to convict me." That's powerfully true, and it's also true of being a conservative. If you're really a conservative, you shouldn't have to tell anybody—you will bear the scars, because you will have been in the trenches fighting.

As the Scripture says, "You shall know them by their fruits." This was the kind of argument that powerfully resonated with the grassroots of Texas.

We also had the good fortune of an overconfident Dewhurst campaign, brimming with highly paid consultants who didn't really understand what was happening with the people of Texas. At the outset of the campaign, with nine candidates in the primary, a variety of grassroots groups, Republican women groups, and tea party groups began hosting candidate forums. I attended almost all of them, as did most of the other candidates. Dewhurst skipped them all.

Given our limited resources, we had to find ways to communicate that did not involve millions of dollars. One of the best ways to do so is with humor. At the cost of about two thousand dollars, we created an animated cartoon, narrated with a Rod Serling–type voice, that began, "Some creatures defy explanation. Elusive objects of great mystery. The chupacabra has been spotted across Texas, a mythical coyote-like animal blamed by ranchers for killing livestock." Across the screen, an animated chupacabra danced in and out of the shadows.

The ad continued, "In East Texas, many have claimed to have seen the tall, mysterious animal known as Bigfoot or Sasquatch. But conservative Texans don't seem to catch a glimpse of the political animal David Dewhurst, despite his height. Since he entered the race on July 21, Dewhurst has skipped nine candidate forums, tea party forums, Republican women forums, and county party forums. He simply won't stand and answer hard questions from grassroots voters." A tumbleweed blew across the screen, past a schoolhouse with a Gadsden Flag.

"Why do Texans have a better chance of spotting a chupacabra than their own lieutenant governor? What is David Dewhurst hiding from?"

If you're going to be at all negative, it's important to be light and funny. Our ad was campy, not vicious. It ended with a picture of Dewhurst dancing across the screen like the chupacabra and Bigfoot.

That video went viral. People liked it. They laughed out loud, and they sent it to their friends. We then ratcheted up that same narrative with a website, DuckingDewhurst.com. Why, we asked, was David Dewhurst ducking the grassroots?

We then purchased a life-size duck suit and would send a young campaign staffer dressed as a duck to Dewhurst campaign events. He would hold a poster board with "DuckingDewhurst.com" written in crayon. It drove the other campaign nuts. They'd have an event with firefighters, hoping for a nice newspaper story about how Dewhurst loves firefighters. But the lead of the story would be, "Today, a man dressed up as a duck greeted Lieutenant Governor David Dewhurst. . . ."

The duck suit started driving the narrative that Dewhurst was out of touch. At a number of the candidate forums, the organizers would set up an empty chair with the name David Dewhurst on it. At one of them, a grassroots activist brought a milk carton with Dewhurst's picture on the side: "Have you seen this man?"

All of it was designed from a grassroots perspective to generate momentum.

For the first phase of the campaign, the Dewhurst camp ignored us entirely. But as we gained steam, they changed course and began to carpet-bomb us. If they couldn't get the voters to love Dewhurst, they figured they'd make them hate his leading opponent.

Dewhurst zeroed in on a couple of the cases I'd litigated in private practice. One involved a trademark dispute between two commercial tire companies. Both of those companies manufactured their tires in China, though one happened to be headquartered in the United States. Morgan Lewis represented the other company, based in China. I didn't actually argue the case, but I was part of the appellate team. Not that such distinctions mattered in politics. The opposing litigant was featured in a Dewhurst ad—one that might have been written in 1950—charging me with siding with the "Red Chinese" against American jobs.

The Dewhurst campaign even printed fake Chinese currency with my face on it, altered to make me look Chinese.

Another attack ad involved litigation connected to a tragic scandal in Pennsylvania in which two corrupt state trial judges took bribes from the owner of two juvenile detention centers. Both the judges and the owner were rightfully incarcerated for their part in the terrible ordeal. Also swept up in scandal was the developer who had built the prisons; he was convicted of failing to disclose the real estate finder's fee he had paid for getting the contract to build the prisons.

Morgan Lewis was retained to litigate the subsequent contract dispute between the real estate developer and his insurance company, and I helped handle the civil appeal.

The Dewhurst camp conflated all of this to assert that I was somehow responsible for the death of children. They ran a heart-wrenching ad with the mother of a wrongfully incarcerated teenage boy who had taken his own life.

The Dewhurst team used their unlimited resources to their advantage. These ads were run to saturation. You couldn't turn on the television without seeing ads about how I supported the "Red Chinese" in killing American jobs, and judges who wrongfully imprisoned children.

When those ads started running, we faced a decision point. At the time, a prominent national pollster was generously giving me informal advice. I sent him the Chinese tire ad. He watched it. I asked him his thoughts. He said, "This is fatal. If you don't respond to this immediately, you're dead." That kind of comment gets your attention. "Fatal" is not a mild diagnosis.

My chief strategist, Jason Johnson, strongly disagreed. He advocated that we not respond to it, because he didn't want the whole race to come down to whether or not I was in fact a "Red Chinese communist." Instead, he advised, we should save our money for communicating our own positive vision. After much thought, I took a deep breath and went with Jason.

One of the data points Jason used to convince me came from nightly tracking polls he suggested we pay for. These are rolling polls taken every single night to see how public opinion is changing in real time. Even though we weren't in an ideal financial position to pay for these expensive polls, Jason convinced me that they were worth the investment. "If you start hemorrhaging because of Dewhurst's attacks," he said, "we'll get a response on. But let's not blow our money first, until we actually see that it's doing real damage."

I'll confess, in those first couple of weeks of the attack ads, I felt a little bit like a boxer who walks into the center of the ring, puts his fists straight up, and just lets his heavyweight opponent slug him repeatedly in the ribs while doing nothing to defend himself.

The tracking polls showed that the attacks were having an effect. My negatives were going up steadily, about a point a night. However, simultaneously, Dewhurst's negatives were going up, about 1.4 points each night. In other words every time he was punching me, it was hurting him even more than it was hurting me. The tracking polls also showed that even though my negatives were going up, the percentage of people who said, "I'm voting for Cruz," didn't dip down.

I thought about these ads later, when I read Senator Marco Rubio's autobiography about his incredible campaign in Florida. Marco, like me, started out as an impossible underdog against the formerly Republican governor Charlie Crist. He ran an inspired, and inspiring campaign. What I found most interesting about his book, aside from the similarity of the campaigns we waged, was the observation that his opponent's campaign was convinced he was an arrogant hothead. Rubio's opponent approached the debates hoping to provoke him into blowing his top.

Marco has become a good friend, and he is many things—but an arrogant hothead is not one of them. His book speculated that his opponent's error was caused by his campaign's assumption that as a Cuban-American, Marco was a fiery Latin. Likewise, the Dewhurst campaign proceeded from the assumption that I too must be a fiery hothead, especially because, from their perspective, it was the height of audacity for someone who was not an elected official to dare run for "their" Senate seat instead of waiting for his turn.

The possibility that I simply believed that our country was in crisis and we needed to get back to the free-market principles and constitutional liberties upon which we were founded never occurred to them. Be that as it may, Dewhurst's effort to prod me into losing my cool proved ineffective.

Under Texas law, if no candidate breaks 50 percent of the vote in a primary contest, the top two candidates go to a head-to-head run-off. In such a situation, the challenger has a significant advantage. When a runoff has been triggered, a majority of voters have already voted against the front-runner. From day one, our strategy was to get to a runoff.

Because voters need to make the time to vote on an unusual election day, runoffs also elevate the importance of intensity, which in our campaign was never in short supply. This intensity began with our campaign staff and was obvious to anyone on Twitter. We paid our staffers about half what the Dewhurst campaign paid their staff, but when the dinner bell rang, Dewhurst's staffers disappeared.

This led to one of the more amusing tweets of the campaign. One activist tweeted: "It's past 6 p.m., so all the paid Dewhurst staffers have gone home. Now all that's left are real grassroots activists, and we're all with Cruz." What was striking was how much truth there was in that sentiment.

Our campaign's strength was that we were blessed with true believers, individual citizens who fought with a passion that was unbridled. If on primary day, David Dewhurst was at 49.99 percent, I was confident we would win a runoff—primarily because of intensity.

But despite all our efforts, to get to a runoff we needed more cash. Two weeks before the primary, I sat down with Heidi and told her, "Sweetheart, I want to put all of our liquid net worth into the campaign."

Now, I don't necessarily advise having this conversation with your spouse. Heidi and I, at that point, had been married eleven years. I was fairly confident she would agree, but I assumed it would be only after a difficult conversation of several hours.

Instead, just a few seconds after I raised the question, she looked at me and simply said, "Yes, let's do it."

Even after having been married for more than a decade, that she agreed so readily left me flabbergasted. For most of our marriage, I'd been in public service; we were living comfortably, but without deep financial resources. Indeed, when I stepped down as solicitor general, we had limited savings and I still had substantial student loans. My years in private practice enabled me to pay off those loans in 2009, and to accumulate about $1.2 million in savings.

We put all of it into the campaign. Had we not done so, in the final week of the campaign we would have been dark on television.

My dad, like Heidi, was an incredible supporter. A septuagenarian who had survived everything from poverty to Cuba's prisons, his energy amazed me. Dad was on the road six days a week, driving alone for thousands of miles, preaching at churches and speaking all over the state.

At one candidate forum in West Texas, my dad appeared as a surrogate for me after driving five hours to get there; he drove all the way back home that same night. When I called to ask how it had gone, he said, "Well, one of the other candidates sent a surrogate as well. At the end of the forum, the surrogate for the other candidate came up and said, 'Can I have a Cruz yard sign?'"

Dad's encounter was a good sign, but outside of our campaign, few people expected the election returns that came in on primary day. To the shock of the political world that had months ago predicted a Dewhurst landslide, David Dewhurst was only at 44 percent. I was at 34 percent. The runoff was on.

In many ways, the runoff was far easier than the campaign had been. Our greatest impediment had been the perception that winning was impossible. The argument made endlessly in the country clubs and Chambers of Commerce across Texas was that Dewhurst was inevitable. In one instant, when the election results came in and we were in a runoff, that argument was shattered. But not with everyone.

As usual, Washington types were the last to get what was happening. The influential online site *Politico* reported after the primary result that "Dewhurst remains the favorite in the run-off. Backed by Gov. Rick Perry and much of his political operation, he will still hold a sizable financial advantage and will be able to tap his own deep pockets at a moment's notice."

Within days of the primary, we were back in the field with another poll to assess where the race was. Our first poll showed us instantly with a double-digit lead over Dewhurst.

It's worth noting how fast those numbers changed. On primary day, we were 10 points down; just a couple of days later we were

15 points up. That's a 25-point swing, in a matter of hours. That's a pretty good measure of the impact of the establishment's "electability" argument; inevitably, they argue that only the most moderate candidate can win. The facts often don't back them up. When their "electability" claim is disproven (as getting to the runoff did in our race), voters often prefer a real conservative and the numbers move dramatically.

Later that week, when Heidi and I were driving to church, she said, "Okay, great. You're in the runoff. We're going to win."

I said, "Yes, that's true. But we've still got to raise another three million dollars in the next three weeks." She nearly had a heart attack.

After she recovered from the shock, however, Heidi stepped up and took the lead, along with Chad Sweet, one of our closest friends, who was volunteering as our finance director. They led our sixty-member finance team in putting together the "60 by 60" project: sixty people raising $60,000 apiece so we could go back up on the air. We told people we were "all in" and that we needed their help.

When I was in Lubbock, Texas, I had an encounter with a voter that was a powerful illustration of the grassroots campaign and the sense that many others were "all in," too. An older gentleman came up to me and grabbed me by the shoulder. He said, "Ted, I'm seventy-three years old. I'm retired. In the primary, I gave you twenty-five hundred dollars out of my retirement savings." Then he told me he was giving me another $2,500 in the runoff, "because if we don't turn this country around, my retirement savings is going away."

It is breathtaking and humbling to look in the eyes of a man who is asking you to help turn this country around, and the fact that it happened repeatedly never detracted from its power. Over and over again, hundreds of times, men and women looked me in the eyes, squeezed my hand as hard they could, and said, "Ted, please, don't go to Washington and become one of them."

Another time, down in the heavily Hispanic and largely Democratic Rio Grande Valley, we had a Saturday night rally with about three hundred people. The chairman of the county Republican Party introduced me by saying, "Most politicians come to the Valley and do a thousand-dollar-a-plate fund-raiser at the country club. Ted is different. This is not a fund-raiser. Ted is here to listen to you, to hear your concerns, and to have conversation."

I got up and said, "Javier, thank you for your hospitality and that very kind introduction. I have to disagree, however, with one thing you said. You said 'this is not a fund-raiser.' *Everything* we do is a fund-raiser. If you can max out and give five thousand dollars, we desperately need the funds. But every one of you can give ten, or twenty-five, or fifty dollars. And we need your help. I cannot win this race. But *you* can. Together, we can win."

By the end of the runoff, astonishingly, we had actually outraised David Dewhurst. With the support of every lobbyist and just about every major donor, he had raised $9 million. We had raised $9.5 million. His came from 3,000 donors; ours came from more than 34,000 donors. Republican women, young people, Hispanics, those men and women at that Rio Grande rally. (Of course, Dewhurst also wrote his own check for more than $25 million on top of what he raised, which is why we got outspent 3 to 1.)

Sometimes people ask me, "When you have a room full of Republican senators yelling at you to back down and compromise your principles, why don't you just give in?" The answer is simple. I just remember all those men and women who pleaded with me, "Don't become one of them." I'm not willing to disappoint them.

By the time of the runoff, Dewhurst's negative campaign was becoming a liability for him. But instead of shifting gears, he doubled down. And because he had so much money, his anti-Cruz flyers flooded mailboxes all over Texas.

I believe strongly in Reagan's Eleventh Commandment: "Thou shalt not speak ill of another Republican." That doesn't mean you shouldn't disagree on policy—campaigns are supposed to focus on records and policy differences—but it does mean that personal and character attacks have no place in politics. And, despite the pounding we took, I deliberately chose not to reciprocate; indeed, to the contrary, I made a point to praise Dewhurst's character and integrity.

And so when one of his flyers went to the wrong address, I asked Dewhurst about it during a debate on live television:

"You know, I think you can tell a lot about a candidate by how they conduct their campaigns. From day one, my campaign has kept the focus on the issues. What we're talking about tonight are amnesty and payroll tax and the lieutenant governor's record and my record. That's what Texans want. Unfortunately the lieutenant governor has not reciprocated in that. He has spent over ten million dollars of his vast personal fortune flooding the airwaves with false personal attack ads, maligning my character.

"You know my dad fled Cuba as a teenager. He was imprisoned and tortured in Cuba and he came here seeking the American dream. Just this week, my father received at his home a mailer from Lieutenant Governor Dewhurst that has the picture of the lieutenant governor on the front of a flag, and an American flag, and then on the back, it has a picture of me in front of a Chinese flag. One of the worst things you can say in politics is to malign someone's patriotism. What the lieutenant governor sent to my father was a mailer that said, 'Ted Cruz worked against our country and lied about it.'"

I turned to the lieutenant governor, who was standing a few

feet away. He was visibly uncomfortable. "You know, I have to say, Mr. Lieutenant Governor, you're better than this. This is not what politics is supposed to be about. This is why people are disgusted with the nasty personal attack ads. What I would ask you standing here today is, do you stand by this? Do you stand by maligning my patriotism?"

At first Dewhurst tried to tap-dance around the mailer, but eventually he began wielding those charges directly at me—accusing me of trying to kill American jobs and supporting the wrongful imprisonment of kids. He was interrupted by boos in the crowd. At that point, it was clear to almost everyone that the campaign was effectively over.

On July 31, the day of the runoff, Texas experienced a record turnout. Even though it was 106 degrees outside, more than 1.1 million voters showed up to vote. And when the votes were tallied, we didn't just squeak by; we won by 14 points.

Indeed, we received more raw votes in the runoff than Dewhurst had received sixty days earlier, when the turnout had been 1.4 million. We were thrilled and somewhat surprised when the Associated Press called the race less than an hour after the polls closed.

Surprised—but not as surprised as David Dewhurst was. That evening, when the lieutenant governor called me to concede, it was clear he was stunned. He had woken up that morning to the news that, according to his pollster, he was up by 5 points. That pollster was off by *19 points*.

In fact, Dewhurst had been told every day of the election that he was going to win, by a team that frankly did not serve his interests. In politics, there are far too many mercenaries who treat campaigns like salesmen selling bars of soap. Their perfect candidate is a candidate with unlimited financial resources, who can afford to keep writing check after check after check. David Dewhurst personally put more than $25 million of his own money into the race. Yet his team did not tell him the truth about the state of the race, presumably because they wanted him to keep writing checks.

The team robbed him blind, not just figuratively, but in fact literally. Months after the election, it became public that Dewhurst's campaign manager had embezzled more than $1 million from the campaign. He was sentenced to 7 years in prison.

I wish I could say I was amazed by this. But so many political consultants are in these races for themselves, not for the people. It is another problem endemic in Washington, D.C.—whose political leaders are captive to consultants very much like the ones Dewhurst retained.

As serendipity would have it, the day of the Republican runoff was also the late Milton Friedman's one hundredth birthday. Having begun studying Friedman at the Free Enterprise Institute nearly thirty years earlier, I could only begin my acceptance speech that evening by noting that it was fitting and perhaps even providential that we could celebrate this victory on what would have been "Uncle Miltie's" hundredth birthday.

Three months later, it was incredibly gratifying to win the general election by 16 points. Not only that, but we had earned broad-based support across the state, including winning 40 percent of the Hispanic vote. Texas is the only majority-minority state in the country that is solidly Republican, and the Hispanic support we received far outpaced the shellacking Republicans were getting nationally.

Indeed, it was hard to celebrate too much that night because Republicans were getting pummeled nationwide. Mitt Romney was losing the presidential race to Barack Obama, and Republican Senate candidates all over the country were losing, too. The only three Republicans newly elected to the Senate in 2012—myself, Jeff

Flake, and Deb Fisher—all won with substantial support from the tea party, a fact often missed by the D.C. pundit class.

In the coming years, the Beltway consultants would continue to insist that the way to win Senate campaigns is to nominate more and more moderate, establishment candidates, even though 100 percent of them lost in November 2012. These were of course the very same arguments used to nominate Bob Dole in 1996, John McCain in 2008, and Romney in 2012. All are good, honorable, decent men, but all three lost. Somehow the fact that you win elections by drawing distinctions, by giving the people a clear reason to show up and vote, has still not penetrated the political consciousness of Washington, D.C.

The cognitive dissonance I had found among the establishment class in Texas was nothing compared to what I would discover in the nation's capital.

★

CHAPTER 8

Into the Beast

The accusation was telling. Only to Washington insiders could building a successful multimillion-dollar business from scratch, raising a seemingly well-adjusted and devoted family, and spearheading a military-style rescue of employees being held in an Iranian prison be considered "inexperience" for national office.*

H. Ross Perot was a man with his own failings, to be sure—he was the first to admit that—but he had built a tremendous business empire from nothing. He found himself on the cover of *Fortune* magazine and was one of the wealthiest men in America.

With pugnaciousness and grit—his major issue was the federal debt—the five-foot-five, silver-haired Perot had managed a Herculean feat: He was running neck-and-neck and in some instances leading national polls for the presidency of the United States against two seasoned political veterans. His third-party run challenged the two-party system in a way it hadn't been since the days of Teddy

---

*The rescue operation inspired a novel and an NBC TV miniseries, *On Wings of Eagles*. Perot was portrayed by actor Richard Crenna.

Roosevelt. As you might imagine, the two-party system didn't like that very much.

In the Athletic Complex on the campus of St. Louis's Washington University, Perot stood beside the Republican incumbent, George H. W. Bush, and his Democratic challenger, Arkansas governor Bill Clinton, for the first of three presidential debates. The two men onstage quickly attacked Perot on the audacity of deciding to run for president when he hadn't in effect paid his dues in politics. "I think one thing that distinguishes [us] is experience," said the sitting president.

Asked to respond, Perot offered a typically forceful, truthful answer that turned the question on its head.

Gesturing to the president and Governor Clinton, Perot shrugged. "Well, they've got a point," he said in his high-pitched Texas twang. "I don't have any experience in running up a four-trillion-dollar debt. I don't have any experience in gridlocked government where nobody takes responsibility for anything and everybody blames everybody else. I don't have any experience in creating the worst public school system in the industrialized world, the most violent, crime-ridden society in the industrialized world.

"But I do have a lot of experience in getting things done. So if we're at a point in history where we want to stop talking about it and do it, I've got a lot of experience in figuring out how to solve problems, making the solutions work, and then moving on to the next one. I've got a lot of experience in not taking ten years to solve a ten-minute problem. So if it's time for action, I think I have experience that counts. If it's more time for gridlock and talk and finger-pointing, I'm the wrong man."

It was a rare feat indeed for a businessman to outshine politicians who talk for a living, but Perot managed it. At one point, the diminutive Texan who was mocked in cartoons for his outsize facial features joked that if someone had a better idea for reducing the deficit, "I'm all ears."

With bracing, often self-deprecating humor and a mastery of statistics and pie charts, he brought to the country's attention a usually arcane topic that most politicians liked to avoid: a huge federal debt created through excessive spending and overpromises by both political parties. By the time the debate was over—Ross Perot was considered its winner by a Texas mile. Perot might well have been the first third-party candidate elected to the White House had he not dropped out of the race, and then in a confusing move dropped back in again.

Perot had also created a true grassroots movements of former Reagan Democrats and Independents fed up with Democrats who believed spending was the answer to most problems and Republicans who seemed to say one thing to get elected and then did another. The roots of what became the tea party movement had its first shoots back in 1992.

———

On January 3, 2013, at 12:10 p.m., I was sworn in to the U.S. Senate. As I stood in the well of the Senate and took my oath of office, I couldn't help but think back to my dad, fifty-six years earlier, washing dishes in Austin. If someone had told that teenage immigrant that five decades hence, his son would be sworn in as a U.S. senator, he would have found it impossible to imagine. And yet, as I stood with my hand on our family Bible, there was my dad in the gallery looking down. Tears were running down his face. It was a powerful moment in our family, and yet one more illustration of the incredible promise of America.

But it didn't take long for it to become clear how different Washington was from Texas. I had been in the Senate for a couple of weeks when I had one of my first instructive encounters with the mainstream media. I was at the annual Alfalfa Club dinner in Washington and the encounter was with NBC's Andrea Mitchell, a doyenne of the Washington press corps and wife of former Federal Reserve chairman Alan Greenspan. As I understood it, Mitchell was a fixture at the countless Washington dinner parties where politicians and media bigwigs dress up in tuxedos and gowns, pal around, gossip, and basically act like governance is just an entertaining game played among the insiders. President Obama is said to hate such events for that reason—it's one of the few things on which he and I agree.

A few days before the dinner I had appeared in a debate of sorts on *Meet the Press* with New York's senior senator, Chuck Schumer, one of the most partisan, quotable, and effective politicians on the Democratic side, discussing the administration's effort to raise the debt ceiling. Schumer took issue with Republicans like me who were urging that the Congress take action on spending restraints before voting again to increase our ever-growing federal debt. He argued that it would be terrible to risk the full faith and credit of the U.S. economy by, as he put it, holding the debt limit increase "hostage."

In the course of our discussion, I turned to Senator Schumer and said happily that we had found an "area of substantial agreement." We should *never* risk the full faith and credit of the United States. Then, right there in front of moderator David Gregory, I offered him a deal. I said he could sign on as a cosponsor of legislation that would effectively remove the issue from the table. He could, I urged, support a bill saying that regardless of what happened on the debt ceiling vote, the United States would always, always, always pay its debt. We could make bipartisan news; together we could permanently guarantee to our creditors that we would never default on our bills.

"I support the concept," Schumer replied haltingly, somewhat flummoxed. To date, he has not signed on to the bill. Nor is he likely to. What he hid from viewers was the fact that Democrats weren't really interested in protecting America's obligation to its creditors as much as they wanted to keep spending on programs to please their constituencies without any limits or reductions.

Indeed, every time a debt ceiling vote arises, Democrats can be counted on to raise the specter of a default on our debt. Of course, no responsible president would ever allow a default. And, regardless of what happens with the debt ceiling, there are ample revenues to service the debt; each month, federal revenues are roughly $200 billion, and interest on the debt is typically $30–40 billion. But the way the Democrats try to avoid spending reforms, and keep racking up trillions in debt, is by threatening a default that, they ominously suggest, could trigger a financial apocalypse. That's why we should pass what I call the Default Prevention Act, which is what I urged Schumer to support. Take default off the table so the debt ceiling can be used to force meaningful spending reforms and actually fix the problem.

After my appearance, Mitchell had gone on the MSNBC program *Morning Joe* to say she was "shocked"—shocked—that I dared confront Schumer. Indeed, she was offended. As she herself later explained, she was not offended that Democrats like Schumer were proposing raising more taxes—"raising revenues" was Schumer's preferred D.C. lingo—on struggling Americans. Nor was she offended that we were doing nothing serious to tackle the debt we were handing to our kids and grandkids. Or that I had made a point against Schumer that he struggled to counter. No, she was offended because she believed I'd violated the Washington rules of decorum. My offer to join together on legislation preventing a default was, she declared, "rude" to poor Chuck Schumer—a man who doesn't strike anyone as easy to bruise verbally.

So at the Alfalfa Club dinner, as she stood next to her husband, Greenspan, I approached the legendary Andrea Mitchell with a smile. "I have to say, you shock quite easily," I said to her.

She looked baffled. It was clear she had no clue what I was talking about. I suppose hers were just words to fit a time slot—an easy slight against a Republican on a liberal news network—and so didn't merit her recollection. But in fact she did remember. Many months later, to buttress her claim that the "Meet the Press" exchange had been "shocking" and "rude," she noted that Schumer had engaged in the usual banal niceties during the segment, calling me "the gentleman from Texas" and "my friend, Ted" even though he didn't know me at all. Most freshman senators, she pointed out correctly, are preoccupied with "making friends" and "going along" with the crowd. I didn't do that, which is supposedly what made me "rude." This fit nicely into the Democratic Party's effort to portray tea partiers like me and my colleagues as brash hotheads who cannot govern a country.

She did have a point in one regard. Words matter. Typically, I like to call people a friend when they've been, actually, a friend. But in Washington everyone is labeled a "friend," especially when you're about to put a shiv in his or her back. Frankly, I doubt whether Senator Schumer really thought I was much of a gentleman, either. His view of constitutional conservatives and probably Texans in general seems substantially less charitable.

I soon learned another lesson in D.C.—this too involved the idea of "friendship" with leading Democrats.

Days after being elected to the Senate, I came to Washington, D.C., for a weeklong orientation. Senate Majority Leader Harry

Reid asked each of the freshmen to visit in his office. By the time I was elected, the Democrat from Nevada had been a creature of the Senate for twenty-four years. A former boxer with a wiry frame, he was not what anyone might describe as overly chatty.

When we met together in his resplendent office in the U.S. Capitol, we engaged in some pleasantries. And then I made a go at being direct. I said, "Harry, you and I are going to disagree on a great many issues, but you have my commitment right now that number one, I will never lie to you. And number two, I will never disparage you personally or impugn your integrity."

Now, in the real world, most people might have responded with something in kind. At the minimum, an acknowledgment. Maybe even a grunt. Not Harry Reid. He just sat there, staring at me with blank eyes behind his wire-framed glasses, the beginnings of a wry smile forming at the corners of his mouth. I couldn't help but wonder if the wily and partisan Democrat was thinking, "What a poor sap! We're going to eat his lunch!"

I had a more pleasant encounter with another prominent Democrat. After my swearing-in in the Senate chamber, we had an unofficial reenactment in the historic Old Senate Chamber. As president of the Senate, Vice President Joe Biden was there, scrappy as always and surprisingly charming. He flirted with my mother, then in her late seventies. When I mentioned to him that Mom had been born in Delaware and had hundreds of cousins there, Biden grinned and said, "Oh, they probably all voted for me." I'll confess my mother laughed back and said, "You're probably right."

The vice president also leaned over to pick up my youngest daughter, Catherine, who was then two years old. She wailed at the top of her lungs, and he replied, "Now, now, Catherine, it's okay. . . . It's a Democrat, but it's okay." We all laughed.

Later that week, that quote from the vice president appeared in the notable and quotables section in *Time* magazine, and Senator Mitch McConnell's wife, Elaine Chao, kindly sent a copy to Heidi and me. When Catherine is married (hopefully many, many years hence) I intend to pull out that article at her wedding and point out that even at age two, she had the good sense to know to scream loudly if a Democrat reaches over and tries to pick you up.

In my early months in the Senate, Republican Leader Mitch McConnell made a concerted effort to befriend me. Although wary, I was glad to reciprocate. In January, Mitch invited me and the other freshman Republicans to join him on a trip to Israel and Afghanistan. It was a productive trip, and particularly meaningful to sit and visit with our soldiers serving in combat. And he invited me to be his personal guest at the Alfalfa Club dinner—where I met with the aforementioned Ms. Mitchell.

When President Obama came to have lunch with the Republican senators that spring, Mitch's staff called my office and asked me to ask the second question of the president, on Obamacare. It was unusual for a freshman to have that opportunity, and I appreciated it.

McConnell gave me all the committee assignments I wanted—Judiciary, Commerce, Armed Services—and a committee assignment I hadn't even asked for—a spot on the Senate Rules Committee, which predominantly consists of more senior senators. He reasoned, correctly, that I would be an ally on the committee against unconstitutional campaign finance reform. Democrats were expected to make yet another run at it, and Mitch and I share a passionate distaste for restricting the First Amendment rights of American citizens. We believe in free speech and know that most congressional efforts to regulate campaign finance spending are far more about protecting incumbents than avoiding corruption.

Shortly after I was elected, Mitch also offered me something that seemed rather unusual for a freshman senator who hadn't even taken office. He asked if I would join the Republican leadership as vice chairman of the National Republican Senatorial Committee (NRSC).

I was surprised, because I had been critical of the Committee's past practice of opposing conservatives in Republican primaries. It had a miserable record—siding with Bob Bennett against Mike Lee; Trey Grayson over Rand Paul; Arlen Specter over Pat Toomey; and Charlie Crist over Marco Rubio. Today, Senators Lee, Paul, Toomey, and Rubio are among the brightest stars in the Republican Party, while two of their vanquished opponents—Specter and Crist—left the Republican Party altogether.

When I mentioned to Mitch that if the NRSC had had its way, every one of those four conservatives would have lost, he promised that the committee would stay out of primaries from here on out. He said he wanted to bring the tea party and the grassroots together with the GOP. I agreed with that goal and based on that commitment— to stay out of primaries—I signed up.

For the first couple of months as vice chairman of the committee, I worked to help the Republican leadership raise money and support. But it soon became clear that the NRSC had every intention of supporting incumbents—in primaries—against conservative challengers across the country. And even in open races, it actively urged donors to give money to candidates opposing tea party conservatives. That didn't sit right with me. I didn't formally resign from my position, but I stopped asking donors to support the NRSC; I didn't agree with what they were doing in primaries, and so I wasn't willing to ask others to fund those efforts. It was yet another lesson: Assurances in Washington come with expiration dates.

Like all freshmen, I spent my first six months in the Senate temporarily assigned to cramped, windowless basement offices. That was fine—I was thrilled to have any office at all—but it revealed something about the glacial pace at which the Senate works. To most folks, it would seem pretty simple math: You have 100 senators and 100 Senate offices. In the private sector, you'd sit down in a conference room and in a couple of hours, assign every office.

Not in the Senate. Historically, each senator was given three days to choose whether to stay in his or her current office or move to another office. After the most senior senator chose, then the next most senior senator had three days to decide. And so on. One needn't be my mathematician mother to figure out that, at that pace, it would take 300 days to allocate the Senate offices. Thankfully, the year I was elected, Chuck Schumer, who was the chairman of the Rules Committee, offered some good news. Now each senator had one day instead of three; thus the process had been condensed to merely 100 days.

The whole time, there were a dozen Senate offices sitting empty, while the wheels of government slowly turned. Frankly, my suspicion always was that it was one part bureaucracy, one part inertia, and one very big part freshman hazing.

More than once, Texans would come to visit their newly elected senator, see me and my staff sharing desks in the subterranean office, and say with a raised eyebrow, "Are you really a senator?"

In my first month in office, President Obama made two important nominations to his foreign policy and national security team. Senator John Kerry, the former Democratic presidential nominee, was to be secretary of state, replacing Hillary Clinton, whose record left much to be desired, and Chuck Hagel, a former Republican senator, would be secretary of defense, replacing Clintonite Leon Panetta. (Panetta, as it turned out, would soon offer a scathing critique of the Obama administration, and the president personally.)

Both Kerry and Hagel came by my basement office to visit with me in what are labeled "courtesy calls" to the senators voting on their nominations. For a man who had publicly labeled the tea party as a threat to our nation and would even demand that the media not give airtime to tea partiers' "absurd notions," John Kerry was surprisingly pleasant and gregarious. This was another D.C. staple—regardless of the scathing things said in public, we are privately all supposed to be buddy-buddy.

And yet he was oddly tone-deaf. His first request was to ask me for my help ratifying the Law of the Sea Treaty—a request that almost caused me to laugh out loud because of my decades-long opposition to treaties that undermine U.S. sovereignty and subject us to the extralegal authority of foreign bodies. It would have been like my asking him to support the full-scale repeal of Obamacare. For a man who prided himself on his foreign policy expertise and who was seeking to be America's chief diplomat, it was puzzling that he wouldn't have understood that a conservative would have serious qualms about the treaty.

Though Kerry was personable, even charming, that was not sufficient reason to support a nominee for America's top diplomatic post—especially one who for nearly four decades had managed to find himself on the wrong side of virtually every foreign policy issue. In fact he was wrong with such stunning regularity that one could almost ask his position and immediately know that the sound position was precisely the opposite. He supported the Sandinistas imposing communist rule in Nicaragua, and vigorously opposed President Reagan's successful efforts to stand up to the Soviet Union—efforts that ultimately won the Cold War. He opposed the Persian Gulf War, fought during the George H. W. Bush administration, which saved Kuwait from Saddam Hussein. He saw the terrorist-supporting Yasser Arafat as a partner in peace. He was sure the 2007 surge in

U.S. troops in Iraq—which rescued that nation from an Al Qaeda assault—was a hopeless strategy. And he lobbied hard to build a new alliance with Syrian dictator Bashar al-Assad. Most of my Senate colleagues didn't seem too concerned about any of that; it was clear from the outset he was going to be confirmed resoundingly. In part this was out of genuine affection for Kerry, a Senate institution. But there was a cynical reason as well—once Kerry resigned his seat, at least a few Republicans believed that Republican Scott Brown would be elected to replace him.

As a general principle, presidents ought to have wide latitude in choosing the men and women who serve in their administration. It was notable that John Kerry, however, did not share that view. During the George W. Bush administration, he had voted against the nomination of Alberto Gonzales as attorney general; against the nomination of Michael Mukasey as attorney general; and against the nomination of Condoleezza Rice as secretary of state. Likewise I did not vote to confirm a secretary of state whose views I believed were harmful to the national security interests of the United States. On that vote, I found myself joined in opposition by only two other senators: John Cornyn, my colleague from Texas, and Jim Inhofe of Oklahoma.

When I went to the well of the Senate to record my vote before the clerk, it drew gasps. At least that's how it felt. It was viewed as somehow impertinent for a freshman senator to dare vote no for such a nomination, especially against an icon of the institution and especially after nearly every other Republican in the Senate, moderate or conservative, had indicated the proper vote I was supposed to make. This was an early sign of many senators' view—that freshmen are supposed to meekly go along with prevailing sentiments. It was markedly different from my view of the office, which is that I had a responsibility to 26 million Texans to do my very best to represent them and to stand for the principles they sent me to Washington to defend.

At roughly the same time, aspiring secretary of defense Chuck Hagel's confirmation process came to a head. When I began to examine the former senator from Nebraska's record, I was astonished. John Kerry's record has been consistently liberal and wrongheaded, and yet Chuck Hagel's record as a senator was markedly to the left of Kerry's when it came to national security. As one of only two senators who opposed renewal of the Iran and Libya Sanctions Act in 2001, Hagel refused to stand up to nations like Iran, whose nuclear ambitions pose grave national security threats to America. And he had repeatedly suggested a moral equivalence between the terrorists of Hamas and Hezbollah, who murder innocent women and children, and the Israel Defense Forces, which defend our democratic ally against terrorists seeking to destroy the innocent.

But Hagel was another nominee we were supposed to back reflexively—even though it was pretty obvious that many of my colleagues who had served with him, especially Republicans, didn't like him and questioned his temperament. I didn't know him personally, but his record was troubling.

At his hearing before the Armed Services Committee, I took the occasion to vigorously question Hagel. Among my questions were inquiries about a video I played of Hagel's asserting that Israel had committed war crimes. He had no good response. All in all, it was a rough outing in front of his former colleagues—and not just because of me. At one point, in response to friendly questioning from Democrats, he mistakenly said the Obama administration's policy was to "contain" an Iranian nuclear arsenal, rather than to prevent Iran from developing one in the first place. Later, he had to be corrected by the Democratic chairman of the committee—on a matter that should be one of, if not *the*, principal concern of the secretary of defense. The consensus, even among his supporters, was that he was ill-prepared for serious questions.

The Senate Armed Services Committee requires nominees for senior positions to submit a disclosure of the income they have received in the last two years. Hagel did so; however, there was a three-year gap between when he had left the Senate and filed annual financial disclosures and when he was nominated to be secretary of defense. Additionally, he had disclosed several substantial payments from financial services firms without identifying the ultimate source of the funds.

Accordingly, I wrote Hagel a letter, asking him to provide more details. Specifically, I asked him to disclose all the income he had received in the last five years, and also to disclose whether any of the payments from financial services firms came directly or indirectly from foreign nations, corporations, or individuals. I circulated that letter at our Senate Republican lunch, and twenty-five other Republican senators joined me, including Mitch McConnell, John McCain, and Lindsey Graham.

These questions were not unprecedented. When Henry Kissinger was named by President George W. Bush to be on the 9/11 Commission, Democrats demanded that he provide a similar disclosure. Kissinger refused, and he withdrew from the committee. Likewise, when Hillary Clinton was nominated to be secretary of state, she voluntarily provided seven years' worth of disclosures.

Hagel's response was astonishing. He didn't follow the Kissinger model of refusing to answer and withdrawing. Nor did he follow the Clinton model of voluntarily handing over the information. Instead he simply refused to answer. In effect, he told the members of the Armed Services Committee to go jump in a lake. And yet he still expected to be confirmed.

My letter had identified seven payments from financial services firms. For six of them, he stated that the source of the funds had not been foreign, without providing details. But for the seventh, a payment of $200,000, he would not offer any such assurance.

When Hagel's nomination came up for a vote in committee, I voted no. I focused on his extreme foreign policy record and also his refusal to answer reasonable financial disclosure questions from twenty-six senators. I noted that his written answer regarding the $200,000 payment—that he could not tell us if it came from a foreign source—raised the obvious implication that it may well have. The answer was important. If, for example, the payment had come from the government of Canada, say for being involved in a dispute over Canadian lumber, nobody would find that troubling for a defense secretary. But, on the other hand, payment from other nations would be highly troubling. Therefore, I observed, "it is at a minimum relevant to know if that $200,000 that he deposited in his bank account came directly from Saudi Arabia, came directly from North Korea."

In hindsight, I made a mistake when I uttered the words "North Korea."

North Korea was on my mind because earlier that day, its government had announced that it was targeting nuclear missiles at three U.S. cities, including Austin, my former home. But the problem with my mentioning North Korea was that there was no credible reason anyone would believe Hagel had received funds from North Korea. His most questionable policy statements had all concerned Hamas, Hezbollah, Iran, and the Middle East—not North Korea. There was no natural nexus or connection.

In uttering those two words, I allowed the White House and the Democrats to change the subject. Immediately I was accused of a "new McCarthyism" by somehow asserting that Hagel had received money from North Korea—an assertion I did not actually make. This is one of the political games played in Washington—to engage in distractions, using smoke and mirrors as needed.

There is an old saying by trial lawyers: If you have the law, pound the law; if you have the facts, pound the facts; and if you don't have either, pound the table. In the game of distraction, the White House, Harry Reid, and the Democrats very ably changed the subject from Hagel's disastrous foreign policy record and his refusal to provide reasonable financial disclosure to the specter of a Texas Republican as the second coming of Joe McCarthy.

No Democrat was willing to defend Chuck Hagel's foreign policy on the merits. Indeed, one of the sorriest aspects of his confirmation was that, despite his repeated history of antagonism to the nation of Israel, not a single Democrat—every one of whom professes to be a strong advocate of our friendship and alliance with Israel—even asked him any serious questions about his extreme record in that regard. Instead senator after senator on the left simply invoked the McCarthy image, trying to end the debate.

Of course, the term *McCarthyism* refers to people making charges for which there is no reasonable basis in fact. In my instance, I was asking for the same financial disclosures that had been asked of Henry Kissinger and Hillary Clinton. And *Hagel's own written answers had explicitly raised the inference that he had been paid hundreds of thousands of dollars from a foreign government*. Given his written admission, any responsible senator would insist on knowing the answer.

We had 45 Republicans, and if 41 of us stood together, we could stop the Senate from proceeding on Hagel's nomination. I worked closely with Mitch, whipping our colleagues to insist on the disclosure. And, initially, we won. Forty Republicans voted against proceeding to the nomination, a remarkable and unprecedented vote against a defense secretary nominee.

But, alas, we could not hold. In the two weeks that followed, the personal attacks intensified. Democrats could not understand how a freshman senator, just weeks into his term, could possibly be leading (and winning) this fight. Ratcheting up the heat, Senator Barbara Boxer of California took to the Senate floor to charge that I had a "list" of names in my pocket, just like McCarthy had. Media outlets lapped it up. MSNBC's Chris Matthews put ominous pictures of me next to McCarthy, growling that we were both "black Irish."

None of the reporters covering the story could be bothered with addressing the substance. Repeatedly, I urged reporters, "for every ten stories you write repeating the personal attacks against me, could you maybe write one on substance? Namely that the nominee (1) absolutely refuses to disclose any of his income for three years, (2) has a long and extreme record of antagonism towards Israel and openness towards Iran's acquiring nuclear weapons, and (3) has admitted in writing that he may have received $200,000 in cash from a foreign government." Actual "news" reporters would want to know, at a minimum, whether our soon-to-be defense secretary had been paid directly by a foreign government. But no such news reporters could be found.

With the media's complicity, the Democrats' political ploy eventually worked. Some two weeks after 40 of 98 voting senators had successfully blocked cloture, the Senate held another vote, and this time cloture was invoked, with only 27 Republicans voting against it. Thirteen of them had flipped sides (including several who had signed the letter with me just a couple of weeks earlier).

Later that day, the U.S. Senate voted to confirm a secretary of defense who had a record of undermining our ally Israel, rationalizing the terrorism of Hezbollah and Hamas, expressing openness

to allowing Iran to build a nuclear weapon, and affirmatively with-
holding information about whether he had received funding from
foreign governments hostile to the United States. It was not a proud
day for the Senate.

The media's personal attacks on me were a sign of things to
come. For a long time, the left has had two caricatures of conservatives: that we are either stupid or evil. I take it as a backhanded compliment that they have, to some extent, invented a third category for
me: "crazy."

Conservatives spend a lot of time complaining about the press,
and it's not always productive. Yes, almost all journalists are Democrats. Yes, reporters have a double standard when it comes to conservatives. But the public is smarter than many elites assume, and time
and again, the people see right through the media's bias.

There is, however, a new, particularly noxious species of yellow
journalism that is beginning to infect what passes for modern political discourse. It's called "PolitiFact." Through this website, left-wing
editorial writers frequently dress up their liberal views as "facts" and
conclude that anyone who does not agree with their view of the world
is objectively lying. Then, left-wing hacks immediately run out and
say, "Look! This conservative said something that PolitiFact calls a
lie. He wouldn't know the truth if it hit him with a two-by-four!"

The first problem with websites like PolitiFact is their heavy selection bias. They pick and choose what to check and what not to
check. For example, in the course of a thirty-minute commencement address at Hillsdale College, I mentioned that Americans had
invented Pong, Space Invaders, and the iPhone. PolitiFact decided
to fact-check that statement, and it turns out part of it was indeed
mistaken: Space Invaders was actually invented in Japan, not the
United States.[1] When they pointed it out, I quickly acknowledged
my inadvertent error, noting that "it sure seemed American when I

played it as a young boy." Because they picked that single sentence, they concluded what I said was "mostly false." Had they picked the paragraph, or the page, or the whole speech, they would have instead rated it "mostly true" (other than a minor error concerning Space Invaders). The rating was entirely a function of the initial selection.

A second, bigger problem is that PolitiFact often labels statements by conservatives "false" because the statements criticize liberals. I recall that PolitiFact labeled as a "pants on fire" lie my statement that President Obama began his presidential administration by going on a worldwide apology tour. In fact, Obama traveled around Europe saying the United States "has shown arrogance and been dismissive, even derisive." It's true that he didn't explicitly add the words "And I'm sorry." But his message wasn't lost on anybody. And unless it is your job to protect the Democratic president and to find fault with statements by conservatives, it's pretty obvious that Obama wasn't exactly bragging about what he called American "arrogance." He was bemoaning it and vowing to do better. In other words, he was issuing an apology.

The third, and biggest, problem is that they regularly define left-wing opinion as an objective "fact." Anyone who disagrees with left-wing opinion is therefore a liar. So, behind a robe of objectiveness and truth-telling, they labeled as "false" my oft-repeated statement that "the greatest lie in politics is that the Republicans are the party of the rich." In the world of PolitiFact, it is *objectively "true"* that Republicans are the party of the rich, even though it is the most vulnerable among us—not billionaires in private jets—who are suffering the most under President Obama's failed economic policies. And it is objectively "true" even though many of the wealthiest donors to political campaigns (the George Soroses and Tom Steyers, not to mention the limousine liberals in Hollywood and Manhattan) donate massively to the Democrats, not to the so-called party of the rich. Their liberal opinions are "facts," and anyone who dares disagree is deemed a "liar."

Another bias in the media is how its members wield the attack of the need to "get things done," and attack those who don't fit their definition of the phrase. We are told we must avoid "gridlock"— another of the mainstream press's favorite words—in the interests of the American people.

We have enormous problems, grave challenges in this country right now. And no one has a greater desire to change the path we're on, and to fundamentally reform this country, than conservatives who have witnessed the eight-year train wreck that has been the Obama administration. We know the violence those policies have done to American families. It is heartbreaking and very real.

But here's another oddity about the D.C. lexicon. "Getting things done" doesn't mean actually fixing the problems we have. It doesn't mean getting anything positive done. It simply means growing government and doing whatever the Democrats want done— and stopping everything else.

For example, when Democrats routinely blocked George W. Bush's nominees, you did not hear members of the mainstream media complaining about Harry Reid contributing to "gridlock" or failing to "get things done." But when a conservative tries to stop disastrous legislation put forward by the left—such as Democratic gun-grab legislation—pundits seem to rush to television cameras to complain that people like Ted Cruz and Mike Lee are blocking progress. We aren't getting things done!

––––––

When the U.S. Senate was first conceived by the Founders, it was meant to be a forum for civilized debate. And for a long time it was, with scholars like Henry Clay, Daniel Webster, Henry Cabot Lodge, and Daniel Patrick Moynihan among its ranks. These were people of ideas who relished a good give-and-take, the clash of intellects, and the possibility of finding common ground. This is not the modern U.S. Senate, where debate is often confused with authoritative Ted Kennedy–style yelling.

Arriving in the Senate, I had been spoiled by my background as a professional litigator, a profession in which matters of import, sometimes matters of life and death, are discussed and debated every day. The rules for a litigator are reasonably simple: You come to the inquiry prepared. You ask questions that require precise answers. And you follow up with questions based on the witness's answers— or evasions. Pointed interrogations are de rigueur in the legal profession, and they usually aren't taken personally.

The Senate is a different beast. Unfortunately, many senators are wholly untrained in this area. Repeatedly, the following pattern plays out at a hearing: A senator reads a question prepared by his or her staff. The witness then dodges the question. And instead of following up with a second and third question based on the witness's evasion, the senator simply moves on to whichever question the staff has listed next. Even a semi-competent witness can bob and weave enough to avoid admitting important truths.

For better or worse, I approach questioning quite differently. That's what I've been trained to do. For example, one time when Attorney General Eric Holder appeared before the Judiciary Committee, I asked him a simple question: "If a U.S. citizen on U.S. soil is not posing an immediate threat to life or bodily harm, does the Constitution allow a drone to kill that citizen?"

"I would not think that that would be an appropriate use of any kind of lethal force," he said.

This was an evasive response. Something can be "inappropriate" without being "unconstitutional." Purposefully, Holder had not an-swered my question—so I tried again.

"With all due respect, General Holder," I interjected, "my ques-tion wasn't about appropriateness or prosecutorial discretion. It was a simple legal question. Does the Constitution *allow* a U.S. citizen on U.S. soil who doesn't pose an imminent threat to be killed by the U.S. government?"

Holder briefly complained about the nature of hypotheticals be-fore again saying that "in that situation, the use of a drone or lethal force would not be appropriate."

Another evasion, so I tried a third time. "I have to tell you," I said, "I find it remarkable that in that hypothetical, which is de-liberately very simple, you are unable to give a simple, one-word, one-syllable answer: no."

Then, for the third time, Holder replied only that a drone killing "would not be appropriate in that circumstance."

Three times I had asked him a simple question with a seemingly obvious answer, and three times he had refused to answer. Only then did he say, with a heavy dose of exasperation, "Translate my [in]appropriate to 'no.' I thought I was saying 'no.'" Perhaps this reflected the view of the left that the Constitution itself imposes no meaningful restraints on government power other than whatever the restraints of propriety might be.

There is a role for drones in military operations overseas. There may also be a legitimate role for drones to act with deadly force to prevent an imminent threat in the United States, like Pearl Harbor or September 11. But there is no plausible argument under the Bill of Rights that the federal government has the authority to use a drone on U.S. soil to target with lethal force a U.S. citizen who poses no immediate danger to the lives of other Americans.

Later that day, my colleague Rand Paul of Kentucky took the Senate floor to begin a filibuster against the nomination of John Brennan as director of the CIA and against the administration's refusal to answer definitely whether the Constitution permits the use of drones to target U.S. citizens on U.S. soil.

Rand is a good friend, and he and I (along with Mike Lee) have charged into many battles together. When Rand began the filibuster, most of our colleagues did not understand what he was doing. They found the inquiry curious, if not quixotic. Indeed, as we sat at the Senate lunch just a few minutes after Rand had begun, several senators expressed puzzlement as to why the question was even being asked.

Before he began the filibuster, Rand had asked if I might go down to the Senate floor and support him. This presented me with a dilemma, because there is a tradition that freshman senators wait a reasonable time before going to the Senate floor to speak. Although it might seem hard to believe today, that was a tradition I had honored and respected. Up until that point in April 2013, I had not spoken once on the Senate floor.

When Rand initially asked, I told him I thought it was too early for me to speak. Just a few months into my term, I was planning to wait at least six months before delivering my maiden speech, out of respect to the institution and its traditions. However, as Rand continued and the fight intensified, I changed my mind.

It occurred to me that on that very day, 177 years earlier, the Alamo had fallen. So, in part because I love Texas history, and in part because Rand was born in Texas, I brought with me to the floor a letter from the Alamo's commander, Colonel William Barret Travis. It was among the great honors of my life to read Travis's letter from the Alamo in my very first speech from the floor of the U.S. Senate.

In that letter, Travis called on Americans "in the name of Liberty, of patriotism and every thing dear to the American character, to come to our aid, with all dispatch." I observed that if the heroes of the Alamo—Travis, Jim Bowie, Davy Crockett—were alive today, I was confident they would be standing shoulder to shoulder with Senator Paul fighting to defend our liberty.

The Senate's rules for a filibuster allow the senator to keep the floor as long as he or she can. During that time, the senator may not sit down, lean on a desk, or leave the Senate floor, even to go to the bathroom. But the senator can yield to another senator for a question. Often, the questions become lengthy, as one senator attempts to provide time for a filibustering senator to rest his voice. In fact, it is not uncommon for a senator with a "question" to rise and speak for thirty or forty minutes and then end his speech with a question like, "Don't you agree?" My question for Rand was whether Travis's letter from the Alamo "gives the senator from Kentucky encouragement and sustenance as he stands and fights for liberty?"

As the night wore on, I returned to the Senate floor multiple times to support Rand's filibuster with lengthy questions. I quoted from Shakespeare's *Henry V*, reading the immortal St. Crispin's Day speech ("we few, we happy few, we band of brothers. . ."). I read from George C. Scott's opening speech in *Patton* (cleaned up a bit to make it suitable for C-SPAN). I even did something that I believe is a first in Senate history: I stood on the Senate floor and read tweet after tweet after tweet that Twitter users had posted in support of Rand.

"Given that the Senate rules do not allow for the use of cellular phones on the Senate floor, I feel quite confident that the senator from Kentucky is not aware of the Twitterverse that has been exploding," I said, holding printouts of hundreds of tweets. "So what I wanted to do for the senator from Kentucky is give some small sampling of the reaction on Twitter so that he might understand how the American people are responding to his courageous leadership."

One of my favorite tweets came from a sixty-three-year-old grandmother. She had never used Twitter in her life, but she signed up that day to stand with Rand. I'm confident that whatever else the future holds, if there ever is a political Trivial Pursuit game, I will be at least an obscure answer to the question of who was the first senator to read tweets on the Senate floor.

The explosion of the "Twitterverse" was an indication that the attention of the American people was riveted on the Senate floor and the fight for liberty. In fact, the people caught on to the significance of Rand's stand faster than many of my colleagues did. A number of them had gone home and settled down for dinner—and perhaps retired to bed—but their staff called them at home and said, "You must get back here. You need to be part of this."

As the night wore on, one senator after another came to the Senate floor to stand with Rand. Some of them weren't quite sure why they were there or what they were doing, but the American people had become so energized that these senators knew enough to want to stand alongside those of us engaged in this battle for liberty.

Unfortunately, with the exception of Oregon's Ron Wyden, no Democrat joined in that parade of supportive senators. I found this astonishing—and disappointing. For many years, Democrats had prided themselves on their defense of civil liberties, and yet one of the saddest realities of the modern Senate is that today's Senate Democrats are far more concerned about standing with their political allies in the White House than they are with defending the Bill of Rights.

After twelve hours and fifty-two minutes, Rand's filibuster ended—in victory. Before he finally yielded the floor, the White House committed to giving him a written answer to the question that he and many of us had raised. The next day, because so many Americans had focused on that fight, the Obama administration was forced to do what it had refused to do for three consecutive weeks: put in writing that no, the Constitution does not allow the U.S. government to use a drone to target American citizens on U.S. soil.

———

On December 14, 2012, a lone gunman walked into Sandy Hook Elementary School in Newtown, Connecticut, and opened fire on classrooms of little children. He murdered twenty children and six adults. It was the deadliest shooting at a grade school or high school in American history. Every parent, especially every parent of young children, could not help but be horrified.

But almost immediately, President Obama acted on his former chief of staff Rahm Emanuel's now-famous instruction: "Never let a serious crisis go to waste."

After the Sandy Hook massacre, the president could have come out and pressed for stronger law enforcement efforts targeting violent criminals and dangerous individuals with significant mental illnesses. Had he done so, the effort would have been met with bipartisan agreement and swift action in Congress. Instead, the president decided to use this tragedy as an excuse to further his long-standing goals of restricting the Second Amendment rights of law-abiding citizens.

In the weeks after Sandy Hook, Democrats were convinced that the president's aggressive antigun legislation was unstoppable. They were sure that their demagoguery, aided by the active political advocacy of the mainstream media, would make standing up to the president's assault on our Second Amendment rights politically toxic. And they knew from long experience on other issues that it was not difficult to intimidate Republican senators into jumping on the train. As it so happened, a different outcome was in store.

The first phase of the battle was the Senate Judiciary Committee, where staunch antigun senators like Dianne Feinstein and Chuck Schumer aggressively pressed for legislation. Schumer could not hide his glee as he rocked back and forth, declaring that mandatory government background checks for every private sale between individual Americans should be subject to government supervision. This, he proclaimed, was the "sweet spot" where the legislation would inevitably land. Meanwhile, Feinstein renewed her push for the so-called assault weapons ban.

The first "assault weapons" ban was passed in the 1990s. It was one of the least effective pieces of legislation Congress has ever passed. After that legislation expired in 2004, the Department of Justice studied its effect and concluded it had precisely *zero* impact on preventing violent crime.

I recall in the middle of the debate, my wife, Heidi, a native Californian who was raised on the Central Coast in a vegetarian family that does not live and breathe politics every single day, asked quite innocently, "Should people really be carrying machine guns everywhere they go?" She was surprised when I told her that fully automatic machine guns have been effectively illegal in the United States for general possession since 1934.

With a confused look, she then asked, "Well, what is an assault weapon then?" I replied that the most accurate definition of an assault weapon under the Democrats' legislation is "any gun that looks scary." The definition has nothing to do with the firing capacity of the weapon. It has nothing to do with the lethality of the weapon. It simply has to do with whether the gun looks like the sort of weapon our soldiers carry into battle.

If, for example, a rifle has a folding stock instead of a solid wood stock, that can make it an "assault rifle." If a rifle has two pistol-grip handles, that can make it an assault weapon. Indeed, if a rifle has even a single pistol-grip handle, under Dianne Feinstein's definition that can make it an assault weapon.

At a Senate hearing, I put up a poster of one of the most popular deer hunting rifles in America. Regularly used by millions of law-abiding Americans, the rifle is not an assault weapon under the terms of the legislation introduced by Senator Feinstein. However, I had with me a $3.95 plastic pistol grip that can be purchased at most sporting good stores. I held the pistol grip up to the rifle and said if you attach the pistol grip, then suddenly this legal weapon becomes an illegal "assault" weapon.

The debate about guns in Washington, like most other debates, is conducted in a fact-free environment. The data about what works and what doesn't, about who commits violent crimes, and about which laws prevent violent crimes, are routinely ignored by Democrats whose real goal is disarming America. If you don't believe me, take the words of the senior senator from California, who stated, "If I could have gotten fifty-one votes in the Senate of the United States for an outright ban, picking up every one of them—Mr. and Mrs. America, turn them all in—I would have done it."

Well, at least Senator Feinstein deserves points for honesty, unlike many of her colleagues on the Democratic side, who pretend that their ultimate goal is not a total gun grab.

Just a few months into my tenure in the Senate, the Senate Judiciary Committee convened a hearing on Feinstein's renewed assault weapons ban, which included a hundred-page list of prohibited and permitted firearms. Sitting on the far side of the panel as the committee's second most junior Republican, I noted that the operative language of the Second Amendment—"the right of the people to keep and bear arms shall not be infringed"—is the same as the operative language in the First Amendment and the Fourth Amendment, which protect "the right of the people peaceably to assemble" and "the right of the people to be secure in their persons, houses, papers, and effects, against unreasonable searches and seizures."

I then asked a simple question of Senator Feinstein:

> Would she deem it consistent with the Bill of Rights for Congress to engage in the same endeavor that we are contemplating doing with the Second Amendment in the context of the First or Fourth Amendment? Namely, would she consider it constitutional for Congress to specify that the First Amendment shall apply only to the following books and shall not apply to the books that Congress has deemed outside the protection of the Bill of Rights? Likewise, would she think that the Fourth Amendment's protection against searches and seizures could properly apply only to the following specified individuals and not to the individuals that Congress has deemed outside the protection of the Bill of Rights?

Though the question was delivered calmly and rationally to a fellow colleague on the Senate Judiciary Committee, a twenty-year veteran of the Senate, the reaction was not what I anticipated.

Senator Feinstein erupted. Defensively, she said, "I am not a sixth grader." She described having seen "bodies that have been shot with these weapons." She professed to have "been up close and personal to the Constitution." She described herself as "reasonably well educated." And then, noting that some guns were exempted by her legislation, she asked in frustration, "Isn't that enough for the people in the United States?"

She did not, however, answer my question. Nor did she even try. Rather, following the very same stratagem that was followed by Democrats in the Hagel nomination, she decided to change the subject and make a personal attack.

I will confess her reaction puzzled me. Of course she wasn't a sixth grader. No one would ask a sixth grader a substantive question of constitutional law. The very fact that I asked her the question demonstrated that I had respect for her knowledge. Moreover, it was quite a surprise to me that on an issue as important as this, she had not anticipated a question like mine. Surely, I figured, she had given some thought to the constitutional right she was infringing. Surely she had some sense that there might be qualms raised about a bill so sweeping.

When I had the opportunity to speak again, I noted that no one doubted the sincerity or passion of Senator Feinstein's antipathy toward guns, but she had chosen not to answer my question. Indeed, she had said nothing about whether Congress can pick and choose which other protections in the Bill of Rights to respect.

Later at the hearing, Feinstein did address this question—kind of. She said, "Congress is in the business of making law. The Supreme Court interprets the law. [If] they strike down the law, they strike down the law."

That answer too was very instructive—and exemplifies one of the biggest problems in the federal government. For far too long, Congress has passed legislation with no one in the Senate once asking what should be the preliminary question: Do we have the constitutional authority to enact this bill?

Feinstein's comment was reminiscent of a comment from then–House Speaker Nancy Pelosi, who responded to a question about the constitutionality of Obamacare by asking in disbelief, "Are you serious? Are you serious?"

Members of Congress don't entertain thoughts about whether

or not their legislation is constitutional for several reasons. For one, they believe and behave as potentates who believe that every crisis, every national headline, demands federal legislation that will impress a subset of their constituents. It's constitutional because they say it is. And if it isn't, then let the Supreme Court deal with it later—long after they've had their press conference and TV interviews trumpeting their great legislative success.

Senator Schumer and Senator Dick Durbin of Illinois both tried to jump in to help their colleague. Schumer airily observed that I just didn't understand what the Supreme Court had decided in *Heller*, the case that confirmed that individual Americans have a right to firearms. I'll admit, I found this mildly amusing because, unlike Chuck Schumer, I had been an actual litigant in *Heller*, representing thirty-one states as amici before the Supreme Court. Indeed, I had presented oral argument before the D.C. Circuit in the companion case to *Heller*.

After I described *Heller*'s holding to Schumer, he went back to a favorite staple of his: discussing child pornography. Whenever anyone suggests that the Constitution prohibits federal authority on a particular matter, Schumer routinely bellows something to the effect of "Child pornography! Rights are not unlimited! We can prohibit child pornography!" His point is that even though watching child pornography could in some people's interpretation be an exercise of free speech, the Supreme Court has upheld legislation prohibiting it. In Schumer's view, because child pornography can be banned, the federal government has carte blanche to violate every part of the Bill of Rights.

What Schumer apparently missed is that child pornography is not an "exception" to the First Amendment that unelected judges created out of whole cloth. For centuries, beginning with the founding generation that ratified the Bill of Rights, obscenity has been understood to be outside the First Amendment's reference to the "freedom of speech." At common law, and in the original understanding, the words simply did not encompass child pornography. That is why it can be prohibited.

When I first asked Senator Feinstein my question, I never imagined it would cause fireworks. It was the kind of ordinary question any lawyer litigating before the Supreme Court would expect to be asked while presenting a constitutional argument. If someone makes an argument that a particular piece of legislation is permissible under the Constitution, the justices will naturally—and predictably—ask whether that same argument would apply to other provisions of the Constitution. That was the world I knew.

But that was not the world I found in the Senate, and I learned this lesson when my exchange with Feinstein became something of an Internet sensation, with nearly a million views online. In the U.S. Senate, senators are not used to actual debate. They are unaccustomed to finding their positions questioned or challenged in any meaningful way. Instead, they expect talking points punctuated by elaborate pleasantries and faux amity, where everyone is a "distinguished gentleman from New York" or an "esteemed colleague from California."

The exaggerated decorum is rendered all the more ironic by the fact that, in the two years I've served in the Senate, this eight-minute exchange with Senator Feinstein constitutes the most words she has ever said to me. Indeed, for months afterward, whenever I ran into her in the elevator, we'd have the following exchange:

"Hello . . . tough guy."

"Diane, I'm all sunshine and smiles."

"Is that . . . what your wife tells you?"

Three times, we've had that exact same exchange, word for word. It's very odd. And somehow, every time she greets me, I hear the words from *Seinfeld*, "Hello . . . Newman."

In any event, the Second Amendment matters far too much for us not to take it seriously. There are two critical reasons.

First, historically, the ability of the people to defend themselves has been a critical precondition to securing liberty from monarchs and tyrants. For that reason, those in government power always want to disarm the populace, because an unarmed populace is subservient to the whims of their masters. When the British Crown became repressive, the American colonists did not throw off the yoke of tyranny by relying on impassioned speeches, pamphlets, or books. Instead they shouldered muskets, launched a revolution, and defeated the mightiest army on the face of the earth. Had the colonists been weaponless, we would likely be speaking with a British accent right now.

Second, the right to keep and bear arms protects the fundamental right of each and every one of us to protect ourselves, our family, and our children from imminent physical harm. If somebody enters my home seeking to injure my wife or my daughters, they will encounter a very direct exercise of my Second Amendment rights.

It's very hard for Hollywood celebrities to understand this. They have no need to own a firearm, because typically they retain armed bodyguards who travel with them. Likewise, other liberal elites living in wealthy suburbs have the resources to hire out their self-protection to the local police through higher property values and property taxes. The threat to their lives, to their property, and to their families is minimized.

But those in poverty don't have that luxury. They are often surrounded by crime, traveling alone late at night, and the Second Amendment is a vital tool to protecting their safety.

That's especially true when it comes to women, who often lack the physical size or strength of a violent attacker. A firearm allows a woman to equalize the playing field, and protect herself and her family.

This was driven home to me when Heidi and I bought our first home in Austin. A small house near downtown, it had one story, with our bedroom window overlooking the driveway. My work sometimes required me to travel, which on occasion would leave Heidi sleeping at home alone. Worried that an intruder might come through the window, I placed a hatchet beneath our bed, and started to tell her to grab the hatchet if anything happened.

As I was saying this, it struck me . . . this was stupid. Heidi is five foot two. The last thing I wanted was for my beautiful, petite wife to be trying to swing a hatchet at a large, menacing robber coming through the window. The next day, I went and bought her a .357 Magnum revolver, which we kept by our bedside.

When it comes to violent crime, I will yield to no one in how strong we should stand for the protection of the innocent. But we do not prevent crime by robbing law-abiding Americans of their constitutional liberties. Instead we target violent criminals and come down on them like a ton of bricks.

It's also worth noting that gun control laws are notoriously ineffective. Facts matter, and cities with the strictest gun control regularly have among the highest murder rates. Thus, D.C. and Chicago have for decades had horrendous crime rates, even though both have been at the extreme vanguard of taking away their residents' gun rights. In contrast, Texas cities like Dallas and Houston and El Paso—where citizens are often armed and able to protect themselves—have murder rates that are a fraction of Chicago's and Washington's.

Similarly, Australia recently followed the path of President Obama and the Democrats, responding to a terrible shooting by banning handguns altogether. The results have been disturbing.

Since banning handguns, sexual assaults and rapes in Australia have skyrocketed, because there are few things a criminal likes better than an unarmed victim.

When the president's gun control bills moved to the Senate floor, the battle entered a new phase. Democrats were still confident they could force through their proposals. But they didn't count on one thing: the power of the American people.

We had seen the impact of the grassroots getting energized in the Rand Paul filibuster on drones. Now we were going to attempt to use the same force to change the balance of power on gun rights. As the Senate was preparing for a two-week recess, I joined with Mike Lee and Rand Paul to send a simple four-sentence letter to Harry Reid. It informed him that we intended to filibuster any legislation that undermines the Second Amendment right to keep and bear arms.

That letter was met with derision by the Democrats, by dismay from the Republican leadership, and by denunciation from their friends in the media. Indeed, the *Wall Street Journal* promptly editorialized how terrible it was that we were making this stand on the Second Amendment.

Our letter accomplished two things. First, it slowed the process down, which gave the American people time to engage. Second, it shined a powerful light on what the Senate Democrats were trying to do.

Over the next two weeks, my Senate colleagues returned to their home states, where they encountered their constituents. Much to their annoyance, their constituents cared deeply about protecting their constitutional rights. One senator after another received angry demands from constituents, who wanted to know why their senator wasn't standing with Rand and Mike and me in fighting for the Second Amendment.

Unsurprisingly, senators did not like being asked that question.

Rarely do elected politicians in Washington expect to be held accountable. Over the course of the next two weeks, one senator after another reached out and asked us to add their name to that letter. Three grew to 17. And then 17 names grew to 31 votes on the first procedural vote against taking up the gun control bills.

Since we needed 41 votes to stop the legislation, the critics immediately declared our effort a failure. But wars, whether in politics or the military, are not usually won in the first skirmish.

Those 31 senators created a whip count—a public record of the 31 senators who would stand with the Second Amendment and the 69 who did not. And after that first vote, Heaven and Earth descended upon those senators who, when at home, tell their constituents they stand for the Second Amendment and yet in Washington, vote against protecting it.

Suddenly the tide turned. As senators began receiving hundreds, then thousands, of calls and letters from their constituents, the political calculus changed. For Republicans, acceding to the efforts of the Democrats to strip away our Second Amendment rights suddenly became a far more risky proposition. And when the day came for the gun control proposals to be voted on, every single one of President Obama's proposals to undermine the Second Amendment was voted down on the Senate floor.

An integral part of defeating the president's antigun agenda was developing a strong, viable alternative. I sat down with my friend Senator Chuck Grassley, then the ranking member on the Senate Judiciary Committee. Together we crafted a bill that became known as the Law Enforcement Alternative. With provisions to increase funding for school safety, reporting requirements for mental illnesses, and prosecution of felons who try to illegally buy guns, the Grassley-Cruz legislation became the key alternative to the president's antigun proposals.

In the course of debating the issues, much of my energy was spent simply describing the baseline facts. As John Adams famously said, "Facts are stubborn things." And there were three facts, in particular, that the Democrats did not want to address.

Fact #1: In 2010, over [48,000] felons and fugitives tried to illegally purchase firearms. Of those, the Obama administration prosecuted only [44] of them. [44] out of over [48,000]. In my view, that is wholly unacceptable. When the Democrats brought in antigun advocates, including police chiefs or U.S. attorneys appointed by the Holder Justice Department, I asked them if it was a top priority to prosecute felons and fugitives trying to illegally buy firearms. Repeatedly, they said yes. Then I asked why in 2010, had they prosecuted only [44] out of [48,000]. They had no good answer. One of the witnesses responded dismissively, "We don't have time to deal with paperwork offenses."

From my perspective, if a convicted felon is trying to illegally purchase a firearm, I want to know why. Rather than having the federal government go after law-abiding citizens, we ought to come down on the murderers and rapists trying to illegally buy guns.

Fact #2: The prosecution of gun crimes under the Obama administration had dropped [30] percent. Whereas the Bush administration had made going after violent criminals who use guns a top priority, the Obama administration put far less emphasis on doing so. And the data demonstrate the consequences.

Fact #3: President Obama reduced the funding for school safety by more than $300 million. That school safety funding might well have gone to providing security to prevent the crazed gunman at Sandy Hook from taking the lives of those little children. But the Obama administration had other priorities.

Each of these facts reflects an ongoing failure of the Obama administration: They have a hard time distinguishing good guys from bad guys. That failure impacts their foreign policy—where we've been ineffective going after terrorists, while at the same time not respecting the privacy rights of American citizens—and it impacts domestic policy, where they've been far more concerned with stripping the Second Amendment rights of law-abiding citizens than with targeting enforcement efforts at violent criminals.

The Grassley-Cruz bill addressed each of these failings, directing law enforcement resources to stop violent criminals from using guns to harm others. It created a gun crime task force, to prosecute violent gun criminals and also felons and fugitives trying to illegally buy guns. It directed resources to helping states report mental health records to the federal background check system. And it enhanced school safety funding, to protect vulnerable children. As a result, it garnered more bipartisan support than any other comprehensive piece of gun legislation—and far more support than the 40 votes Dianne Feinstein's so-called assault weapons ban received. With votes from 52 senators—9 Democrats and 43 Republicans—Grassley-Cruz could have become the law of the land—if Harry Reid and his Democratic allies had not filibustered it.

Reid and the Democrats didn't object to any of the bill's provisions. None of them made a single argument on the merits. Instead, they made clear that the entire post-Newtown gun debate had all been a political exercise. It had been an effort to get their big-money donors excited to write checks to the Democratic coffers. Harry Reid's view was "If we can't get what we want, nothing will pass the United States Senate."

On the evening we were voting on these proposals, Vice President Joe Biden made an appearance on the floor of the Senate. Although he's president of the Senate, Biden rarely appears on the floor. His vote matters only if there's a 50–50 tie, which has not occurred in the time I've been in the Senate. But he was there that night because he was prepared to gavel in from the president's chair what he saw as a historic victory for the left, stripping away the Second Amendment rights of the American people.

Earlier that week, Biden had been calling senators, trying to whip votes. He was reported to have said repeatedly, "Why on earth do you care what a freshman senator is saying?" But he didn't understand; it wasn't what I was saying. That didn't matter. What matters is what happens when the grassroots holds elected officials accountable.

Before the votes began, I was standing on the Democrat side of the chamber, visiting with Senator Joe Manchin. Manchin, a West Virginia Democrat, had pushed hard for one of Obama's gun control provisions. He's a good guy—genuinely likable and down-to-earth—and I was complimenting Joe on how he had conducted the debate, taking the high road and not engaging in personal insults or attacks. While we were talking, Biden came up behind us and rested his hands on my shoulders. He turned to Manchin and said ruefully, "Joe, there's nothing more dangerous than a smart Republican."

As it turns out, Biden's vote was not needed that night. The American people had risen up, and engaged mightily. And not a single bill to limit the gun rights of law-abiding Americans passed the Senate. Democrats were shell-shocked. They had expected an easy victory. But the Second Amendment had once again survived demagoguery and fearmongering, for one reason: the power of the grassroots.

★

CHAPTER 9

The Obamacare Battle

At forty-two, with close-set eyes and a full thick brown beard, Sherman Booth was one of the most prominent organizers of a fledgling political party in Wisconsin. They called themselves Republicans. He was an organizer, a troublemaker, a deeply passionate adherent to principle, and in 1854 he was under arrest.

The scourge of slavery was a deep-rooted vexation to the former Yale student, who during his time at the university had taught English to some of the African slaves who had forcefully taken over the slave ship *Amistad*. A prolific writer, Booth had regularly pressed his case against slavery ever since, most recently in the pages of the *Milwaukee Free Democrat,* of which he was the publisher.

Prominent in his pages these days was the case of Joshua Glover, a Missouri slave who'd escaped from captivity and sought refuge in the Wisconsin city of Racine, known to be friendly to the abolitionist cause. When Glover's slave owner learned of Glover's whereabouts, he filed suit to recover his "property" under the aegis of the notorious Fugitive Slave Act of 1850.

That law, passed as part of the Compromise of 1850 to settle growing divisions over slavery across the young nation, allowed slave owners to demand the arrest, seizure, and return of escaped slaves. In March 1854, acting under the authority of the disgraceful law, a group of men that included a federal marshal tracked Glover down and took him to a Milwaukee jail to await his extradition.

Learning of the capture, Sherman Booth visited Glover in prison and then, on horseback, rallied abolitionists to a protest in the court-house square. As thousands gathered in support of the captured slave, Booth announced that Glover's lawyers' efforts to free him were failing. Knowing well the penalty for their action, a large group of men stormed the jail. Glover was liberated from his cell and then escorted back to Racine, where he made contact with the Under-ground Railroad and fled to Canada.

"In Wisconsin the Fugitive Slave Act is repealed!" Booth exulted in his newspaper. "The first attempt to enforce the law in this state has signally, gloriously failed!" Four days later, Sherman Booth was arrested, facing charges for his willful incitement to violate fed-eral law.

Undeterred, Booth told his inquisitors in court that he stood behind his call for accused fugitive slaves to have trials by jury, a provision directly forbidden by the act. He maintained his opposi-tion to the abominable federal law and said he was willing to face the consequences.

Booth's principled, impassioned determination to use any and all means to fight against an unjust and unconstitutional law led to a rare and memorable standoff between a state and the federal government that captured the nation's attention. The state courts in Wisconsin refused to prosecute Booth, while a federal judge, in a separate trial, instructed jurors to find Booth guilty.

Ultimately, in 1861, President James Buchanan decided to rid the nation of a case that continued to divide the citizenry, issuing a full pardon of Booth for his role in the Glover escape. The pardon was issued just one day before his successor took office—a man whose involvement in the divisive issue of slavery was just beginning: President-elect Abraham Lincoln. He was the first president elected to office under the Republican Party's new banner—a party established to oppose slavery and defend human freedom.

To this day, Samuel Booth stands as one of the great champions of human freedom. He was an American hero who believed that he owed allegiance to the Constitution and brooked no tolerance for laws that in his view perverted the rights God had granted to all of the people.

In my time in Washington, no battle has consumed more energy than stopping Obamacare. On the evening of September 24, 2013, it began with a prayer.

In my tiny "hideaway" office wedged into a dome in the Capitol Building, Senator Mike Lee and I bowed our heads, read from the Book of Psalms, and asked for the Lord's guidance. I then walked to the floor of the U.S. Senate and announced, "I intend to speak in support of defunding Obamacare until I am no longer able to stand."*

I opened by noting that "all across this country, Americans are suffering because of Obamacare." And yet politicians in Washing-

---

* Some pundits quibble over whether my speech was technically a "filibuster." A traditional filibuster cannot be stopped by another senator, because the speaker has control of the Senate floor. But Majority Leader Harry Reid had the floor locked up under preexisting time agreements. It could be weeks before the floor was open and a senator could rise unimpeded. And time was of the essence: Obamacare was set to kick in on September 30. We had to impact the debate before then, so I took the floor pursuant to an agreement that Reid would get it back no later than noon the next day, a little more than twenty-one hours hence.

ton were not listening to the concerns of their constituents. They weren't hearing the people with jobs lost or the people forced into part-time work. They had no answers for the people losing their health insurance, or the people who are struggling.

With good reason, men and women across America believe that politicians get elected, go to Washington, and stop listening to them. This is the most common thing you hear from the man on the street, from Republicans, Democrats, Independents, and Libertarians: *You're not listening to me.*

The fight over Obamacare was an effort to change that. At the outset of my filibuster, I observed that "I hope to play some very small part in helping provide that voice for them." And over the course of the next twenty-one hours, we used a new hashtag, #MakeDCListen, to help Americans communicate with those inside the Beltway. To date, it has received over a million mentions on Twitter.

Before the filibuster began, I received some helpful advice from Rand Paul, who had recently delivered his own thirteen-hour filibuster on the use of drones against U.S. citizens. He said, "Number one, wear comfortable shoes. Number two, don't drink a lot of water."

Rand didn't follow his first piece of advice, because his filibuster had been unplanned. He said his feet hurt for two weeks afterward! But his recommendation put me in a bit of a moral quandary. Every day on the Senate floor up until that point, I had worn my argument boots—the same black, ostrich-skin cowboy boots I had worn before courts in Texas and before the U.S. Supreme Court.

But even those boots can be uncomfortable when worn for that many hours in a row, so I went and bought a pair of black tennis shoes. During the filibuster I confessed to the American people that, to my great chagrin, I had taken the wimpy way out. But as the hours wore on and my legs began to cramp up, I was grateful for the tennis shoes.

I also followed Rand's second piece of advice; as he admitted later, his filibuster was defeated not by his legs but by his bladder. Senate lore suggests one alternative route, which was taken by the late Strom Thurmond in 1957 during his record-setting twenty-four-hour filibuster: Thurmond allegedly walked to the back of the Senate chamber and relieved himself into a bucket, while his feet remained planted on the Senate floor. There are many reasons why I followed Rand's advice and drank only one small glass of water over the course of twenty-one hours, rather than following in the footsteps of Senator Thurmond. Among those reasons is C-SPAN.

Indeed, the most frequent question I am asked about my filibuster is whether I had special provisions that enabled me to use the bathroom. Let me just say, having drunk only one small glass of water, I simply did not need to go.

When you stand on your feet for twenty-one straight hours in defiance of the president, the Democratic Party, and even many in your own party, you find out who your friends are, who your adversaries are, and who's in between.

Over the course of the filibuster a number of allies stood by my side, none more courageously or indispensably than Senator Mike Lee. Mike was with me on the Senate floor during the entire course of the filibuster, from the beginning, through the dark of night, till the very end at high noon the next day. He repeatedly asked me questions that would fill twenty, thirty, or forty-five minutes to give my voice a rest so that I could resume the filibuster. Other supporters included the steady Kansan Pat Roberts; the fiercely conservative Alabaman Jeff Sessions; the Wyoming stalwart Mike Enzi; the Iowa senior statesman Chuck Grassley; and my friend and fellow Cuban-American Marco Rubio.

Another tea party senator was notably less helpful. My friend Rand Paul came to the Senate floor to ask questions that seemed deliberately designed to undermine our efforts. He asked, "Do you want to shut down the government or would you like to find something to make Obamacare less bad?" And, "Will you accept a compromise? Will you work with the President?" His questions echoed the skeptical attacks of Mitch McConnell, and I marveled that Rand had decided not to be with us in this fight. Mike Lee is not an easily excitable guy, but he was so upset by this that I thought he was going to need a sedative.

One of the most memorable aspects of the filibuster concerned my daughters. My two girls, Caroline and Catherine, are the loves of my life. They are precious balls of joy. Our family lives in Houston, and I commute back and forth every weekend. The single hardest aspect of my job is being away from them every week. It breaks my heart on Monday mornings when I walk out of the house and one girl grabs one leg, one girl grabs the other, and they say, "Don't leave, Daddy."

Each night that I'm home, after they put on their pajamas and are ready for bed, I read them a story. So during my filibuster against Obamacare, I took the opportunity to do something I don't usually get to do from Washington, D.C. At 7 p.m. Central time, right at their bedtime, I used C-SPAN to read them a pair of bedtime stories, before telling them good night.

The first was a Bible story. The second was Dr. Seuss's *Green Eggs and Ham*. The delighted girls, in their pink PJs, got right up next to the television and giggled with glee.

Now, my oldest daughter, Caroline, is a bit of a cynic. Absolutely nothing I have done in the Senate has impressed her . . . except for reading *Green Eggs and Ham*. When I came home afterward, Caroline looked at me with crossed arms. "Okay, Dad," she said, "that was kind of cool." We all enjoy small victories in life, and doing something that my then-kindergarten-aged daughter thought was "kind of cool" meant a lot to me.

To this day, people all over the country bring me copies of *Green Eggs and Ham* to sign for them. When I was a child, it was my favorite story—the first I learned to read. (Actually, my mom was convinced I didn't read it but simply memorized the words.)

It is not surprising—or a bad thing—that among the parts of the filibuster that garnered the most attention were the bedtime stories, a *Star Wars* imitation, and a reference to a fast-food chain. The purpose was to highlight the harms caused by Obamacare, and by using humor, my remarks got heard by people who would never otherwise have listened. When Jon Stewart ridiculed them mercilessly (and hysterically), he also included a fair amount of the substance, that Obamacare was costing people jobs, health care, and their doctors. Likewise, the host of *Saturday Night Live*'s "Weekend Update" joked the next weekend, "Texas Senator Ted Cruz gave a twenty-one-hour speech on the floor of the Senate during which he read Dr. Seuss's *Green Eggs and Ham*, did an impression of Darth Vader, and admitted his love for White Castle. I'm not sure what Cruz's speech was arguing for, but I'm guessing legalizing weed."

———

Nearly five decades before there was a tea party movement to oppose Obamacare, there was a coffee party movement. It began in the early 1960s when doctors rallied together to try to save their profession and their patients from a federal takeover of the medical system.

The doctors' strategy was to enlist the most powerful allies they knew: their wives. Each of them would invite a group of friends over for coffee to hear a vinyl record with a ten-minute speech on it. The voice on the record belonged to the host of a weekly television drama called *General Electric Theater*. His name was Ronald Reagan.

Toward the beginning of the record, Reagan noted that "most people are a little reluctant to oppose anything that suggests medical care for people who possibly can't afford it."[1] But he explained that many of the people exploiting the public's goodwill were really searching for "a mechanism for socialized medicine capable of indefinite expansion in every direction until it includes the entire population."

Reagan understood that there has always been a persistent embrace on the part of the left for socialized health care. It is for them the Holy Grail. But they are sailing into the winds of a freedom-loving public that knows that in every country where government takes over medicine, it has meant poor quality and fewer doctors. Socialized medicine has been tried around the world. And from Cuba to Great Britain, everywhere it has been tried it has proved an abject failure. Inevitably, it means low-quality care, rationed by bureaucrats getting between you and your doctor. Scarcity. Waiting periods. Denying individual choice and freedom is antithetical to the American way and our tradition of liberty.

At times (as in Canada today) socialized medicine means waiting months or years to get an operation, because there are thousands of people ahead of you on a waiting list. At other times (as in France today) it means being told that your advanced age precludes you from receiving a hip replacement, because the government has decided that walking is for the young.

Because the American public never did—and, I believe, never will—voluntarily choose the scarcity and rationing of socialized medicine, its proponents have always resorted to trickery, telling lies about the legislation they introduce and hiding their agenda for expanding it. For that reason the left doesn't call it "socialized medicine." They prefer the more innocuous, sterile term of a "single-payer system." The "single-payer" is of course the government, which would then pay every doctor and health-care provider in the country. And when the government pays, it decides: what health care you receive; which doctor you see; how much is a fair price for a product or service in an elaborate scheme of wage and price controls.

For nearly fifty years after the release of "Ronald Reagan Speaks Out Against Socialized Medicine," Americans fought back against the statism that Reagan feared. Jimmy Carter's efforts failed in the 1970s. So did Hillary Clinton's in the 1990s. But in 2010, despite overwhelming opposition from the American people, President Barack Obama found just barely enough support in Congress, both houses of which were controlled by Democrats, to pass Obamacare.

All of us remember President Obama looking at the television camera and saying, "If you like your health insurance plan, you can keep it. If you like your doctor, you can keep your doctor. Period."

This promise provided just enough political cover for Democratic legislators with moderate or conservative constituencies—like Arkansas and Louisiana—to vote for Obamacare. It was a promise that Obama repeated twenty-eight times while looking directly at the television cameras. And it was a promise that was flatly and unambiguously false. Indeed, it was in direct conflict with Obamacare's purpose and plain text, which said that millions of preexisting policies were going to be rendered illegal.

In its perpetual role as the praetorian guard of the Obama presidency, the *New York Times* has helpfully explained that President Obama simply "misspoke" when the falsehood was revealed. But one does not *misspeak* twenty-eight times. Nor is it "misspeaking" to deliberately and repeatedly state something that you know to be false.

Jonathan Gruber, the self-described "architect" of Obamacare, later openly admitted that the law never would have passed if President Obama had told the truth about the fact that what was labeled a "penalty" for noncompliance was actually a tax. Gruber explained what the Obama team was cynically counting on: "And, basically, call it 'the stupidity of the American voter,' or whatever, but basically that was really, really critical to getting the thing to pass."

The edifice of lies on which Obamacare was built has by now been exposed at the expense of millions of Americans who have lost their health insurance, beginning with customers in the individual markets. More than six million of them have seen their health insurance taken away, have been informed that they can no longer see their doctors, or have seen their premiums grow by as much as 60, 70, and sometimes over 100 percent. In Texas, for example, a healthy, twenty-seven-year-old man in the individual market has already seen his premiums grow by more than 77 percent.

The next people to feel the brunt of Obamacare are those in the small-group markets. In the coming years, more and more of the small-group clients will find their health insurance plans canceled and will be informed they can no longer see the doctors they have grown to trust.

The final shoe to drop will be the pain inflicted on Americans employed by large companies. Roughly 100 million Americans get their health insurance from large-group employers, and as the costs of Obamacare rise and rise and rise, we will see more and more large-group employers cutting their health insurance and dumping all of their employees into the public exchanges, where they will suffer along with everybody else.

You might be wondering whom, if anyone, is Obamacare even *supposed* to benefit? Well, all of this was ostensibly designed to extend health insurance to some portion of the people who didn't have it. Not all of them, mind you. The Obama administration concedes that Obamacare will still leave some 26 million Americans without health insurance. But, the estimates go, in time as many as 14 million new people will get some insurance.

That's a good thing. But how would the American people have reacted if President Obama had said, In order to extend insurance to 14 million people, we're going to jeopardize the health insurance of about 200 million Americans? Many of you will lose your policies, and your doctors. Others will see your premiums skyrocket. That is the absurd bargain at the heart of Obamacare. That's why Gruber noted, rightly, that it could be passed only through deception.

You might also be wondering how the Obama administration thought they would get away with this disaster. I think their intention was simply to blame insurance companies when people started seeing their health insurance plans canceled. Liberals excel at vilifying the business sector, and the more they can demonize private-sector insurers, the more leverage they believe they will have for continuing to move toward the Holy Grail of the left that Ronald Reagan warned against in 1961—a single-payer, government-funded, socialized health-care system.

There is a rich irony here, in that the large insurance companies eagerly embraced this radical agenda. They were lured into bed with Obamacare by the government's promise to force every American to buy the insurers' product, which would then be subsidized by the government. It reminds one of Lenin's famous observation: "The capitalists will sell us the rope with which we will hang them."[2]

Obamacare is not only responsible for costing millions of Americans their health insurance. It has also cost millions their jobs. In an era in which 93 million Americans are not working due in large part to a host of job killers authored by Barack Obama, Obamacare is the biggest job killer of them all.

Time and again, I have sat at small business roundtables in communities all across Texas. I generally begin by asking the small business owners to introduce themselves and to share an issue that is weighing on their hearts. Invariably, and without prompting, more than half the people there will name Obamacare as the single greatest impediment their businesses are facing.

At one roundtable in Kerrville, a beautiful town in the Hill Country of central Texas, we met in a restaurant and bar owned by a husband and wife. They described how they recently had a terrific opportunity to grow their business, to nearly double it. From a business perspective, they thought it made a lot of sense. But they'd already passed on the opportunity. The reason? They currently employed between 35 and 40 people, and if they grew to more than 50 employees, they'd fall under Obamacare. And that would bankrupt them.

At that particular roundtable, the first four participants *all* described the exact same circumstance. Each of them had successful small businesses. Each had 30 to 40 employees. Each had significant opportunities to create more jobs. But if they expanded, their businesses would then become subject to Obamacare's employer mandate, and with that extra cost, they wouldn't be able to stay in business.

The people hurt most by that policy were not the men and women at the roundtable in Kerrville. As small business owners, they were struggling, but still able to make ends meet. Instead, the people suffering the most were the people they were not employing—the single moms, the teenagers in community college, and newly arrived immigrants looking for work they can't find. Countless entry-level jobs—for dishwashers and busboys, for administrative assistants and custodial workers—simply don't exist because of Obamacare.

Small businesses have an enormous impact on the American dream, because *two-thirds of all new jobs come from small businesses.* And, under Obamacare, millions of small businesses are not creating millions of new jobs.

At the table in Kerrville was another gentleman, who manufactured hunting blinds. He described how he recently moved his entire operation overseas to China. "I'd like to manufacture these blinds here in the United States," he said. "That would be a hundred and fifty to two hundred good-paying manufacturing jobs right here in central Texas. But if I manufacture in America, the company would be subject to Obamacare. We can't stay competitive in the market; we can't stay in business under Obamacare."

Finally, over to the right of the table was a woman who owned several fast-food restaurants. She already employed well over 50 employees and so staying below that threshold was not an option for her. Instead, she described how she has already forcibly reduced the hours of her employees to 28 or 29 hours a week because Obamacare kicks in at 30 hours per week.

At that point, her voice softened, and she started to choke up. "Many of these employees have been with us for five and ten years," she said. "These are single moms, and these are people who are struggling. They can't feed their kids with twenty-eight to twenty-nine hours of work a week. . . . But they can't feed their kids if I go out of business, either."

These are the voices Congress isn't listening to. When I was listening to their stories in Kerrville, I couldn't help but think about their challenges from the perspective of my father in 1957, washing dishes for fifty cents an hour. If he had been working today, the odds are high that he would have lost his job because of the $1.7 trillion in new taxes and crushing regulations like Obamacare hammering small businesses. And if my dad had been lucky enough to keep his job, the odds are overwhelming that he would have had his hours forcibly reduced to 28 or 29 hours a week.

You can't pay your way through college like my father did on 29 hours a week. What the federal government is doing right now is yanking up the ladder of opportunity for millions, making it harder and harder for those who are struggling to achieve, or even imagine, the American dream.

I was reminded of this unfortunate reality in the spring of 2014, when I attended a large rally of roughly a thousand people in North Platte, Nebraska. A young woman came up to me, hugged me, and said, "Ted, I'm a single mom. My husband left me and won't pay child support. I've got six kids at home, and I'm working five jobs. Not a single one of the jobs is even thirty hours a week, because Obamacare kicks in at thirty hours a week."

The woman in North Platte told me about her struggles to keep clothes on the backs of her children, then added that the hardest thing for her is that she hardly ever gets to *see* her children. "I go from one job, to another job, to another job," she said, her voice trembling, "and I don't get to be the mom that my kids need me to be."

Five years ago, when Obamacare was being debated, reasonable minds might have differed about whether it would raise costs, re-duce quality, and kill jobs. But today, when we have seen firsthand its devastating impact on people who have lost their insurance and entry-level workers who have lost their jobs, the effect of Obamacare is no longer open to debate. American history is replete with mis-guided experiments and social engineering, and yet the disaster that is Obamacare stands unique in its scope and devastation. Today, given the suffering it has caused, the most reasonable, pragmatic approach is to acknowledge it isn't working, repeal it, and start over.

So what was our plan to stop Obamacare? In Washington, pundits repeatedly intone that we had no plan, no strategy, and no hope of success. No doubt, I've got many personal faults, but, as a former Supreme Court litigator, failing to plan is not one of them. We had a systematic strategy, and I'm convinced that the fight we waged in the summer and fall of 2013 accomplished a great deal.

Beginning in the spring of that year, Mike Lee and I began asking our Republican colleagues, "What are we going to do to stop Obamacare from kicking in?" And week after week, they gave the same answer: nothing. Risk aversion dominated their thinking; a fight was risky, and could imperil reelection. Therefore, Obamacare was going to go into effect on September 30, and no Republicans had any strategy to do anything about it.

That answer was not acceptable to Mike and me, nor to the men and women who elected us. And so we suggested an alternative strategy: When the continuing resolution funding the government expired in late September, Congress should fund the entire federal government—but not Obamacare. This strategy was based on the principle that the Constitution is designed with checks and balances. The most significant check that Congress possesses is the power of the purse, and so we urged that Congress should continue to fund everything in the government—except Obamacare.

Our strategy of employing Congress's power of the purse entailed a four-step plan for success. Step one was to take the case to the American people and mobilize millions of Americans against Obamacare. Step two was for the Republican majority in the House of Representatives to do the right thing by passing legislation funding the federal government but not funding Obamacare. Step three was for Senate Republicans to join with House Republicans to pass the same laws. And step four—the final step—was to systematically pick off red-state Democrats in the Senate, one at a time, until we had sufficient numbers to succeed and send the bills to President Obama's desk.

This was, of course, the same strategy we had deployed—successfully—in the fights over drone strikes and President Obama's gun control legislation: empower the grassroots to change the rules in Washington. But here the stakes were higher, and winning more difficult.

Mike and I were never Pollyannaish about what was possible about this plan. Nor did we ever promise anyone that it would be easy. In particular, we thought very hard about one critical obstacle and how to overcome it.

That obstacle was President Obama's veto power. Because of it, the chances of *fully* defunding Obamacare were not high. Possible, I believed, but it would have taken a perfect storm. However, there were a lot of middle outcomes, where, if we united Republicans and put enough heat on red-state Democrats, Obama might feel enough political pressure to agree to some sort of compromise. Not everything, but a middle ground that at the very least gave some material relief to the millions suffering under Obamacare. (And, if you want to end up at a middle ground, you don't start the negotiation by surrendering everything at the outset.)

When Mike and I explained this plan to our Republican colleagues, their reaction was immediate, visceral, and virtually unanimous. "Absolutely not!" "A terrible idea!" Indeed, they openly laughed at the idea, telling Mike and me that we just didn't understand how Washington works.

But when we asked them for their alternative, there were crickets. There was no alternative—do nothing was leadership's alternative—but they vehemently opposed our plan. "Wait until the debt ceiling," they advised. (Of course, when the debt ceiling did come along, their plan was once again . . . to do nothing.) Doing nothing was not an option.

We believed in the power of the grassroots—and that we should fight Obamacare from the outset—and so Mike and I started with step one. Together, we traveled the country barnstorming at rally after rally, where thousands upon thousands of Americans stood together to stop Obamacare. And the results were astonishing. More than two million Americans ended up signing the national petition at DontFundIt.com to stop Obamacare. And those millions then lit up the phone banks on Capitol Hill, calling their members of Congress to say stand up, lead, and do not fund Obamacare. Our colleagues had scoffed that the grassroots would rise up, but, by any reasonable measure, step one of the plan succeeded beyond all expectations.

Step two succeeded as well when, to the wonder of D.C. pundits, Republicans in the House listened to their constituents. Again, our Senate Republican colleagues had laughed that summer, saying there was no way on earth that the House would vote to defund Obamacare. Leadership didn't like the plan. But, for a body reelected every two years, getting more than two million calls from your constituents has a powerful way of focusing the mind. House Republicans stood together to fund the federal government in its entirety and not to fund Obamacare.

Alas, step three was where the plan went awry. It depended on Senate Republicans standing together in support of House Republicans to defund Obamacare.

My assumption had been that the Senate Republican leadership would behave at this step as they had behaved on almost every other significant recent battle. I anticipated that they, and every Republican up for reelection in 2014, would vote with the conservatives, only because it was manifestly in their political self-interests to do so. Then, *quietly behind the scenes*, they would try to pick off 6 Republicans and urge them to join with the 54 Democrats so that our fight wouldn't succeed. That was the Republican leadership's modus operandi. What we did not anticipate was that Mitch McConnell and the GOP leadership team would decide to publicly, directly, and aggressively lead the fight against the House Republicans and in favor of Obamacare.

Perhaps they wanted to discourage conservatives like Mike and me from ever again rebelling against the party line. Or perhaps they were simply angry that a handful of senators would have the temerity to take our case straight to the American people. But for whatever reason, the Senate Republican leadership decided to direct their fire not on Democrats or on Obamacare, but on conservatives in their own party. The result was that, in unison, around twenty Republican senators went on television, on every single news channel, carpet-bombing the House Republicans—and us.

Why then, the critics said, did you shut down the government, if you didn't have the support of the Senate Republican leadership? Two answers: One, if you wait for Senate Republican leadership, we will *never* stop Obamacare. At every stage, their plan was to avoid the fight.

And two, *we* didn't shut down the government. Neither I, nor Mike Lee, nor the House Republicans even once voted to shut down the government. To the contrary, over and over again we voted to fund the government.

From the outset I stated repeatedly that we should fund the entire federal government and defund Obamacare. But after every vote in the House to fund the government, Harry Reid and the Senate Democrats said: We don't like your legislation; therefore we're going to shut down the government. They not only refused to vote for the House-passed bills that funded the government; they refused to even discuss or negotiate any possible relief for the millions being hurt by Obamacare.

Senator Reid's conduct was politically understandable. He *wanted* a government shutdown. He said publicly he thought it would help Democrats politically. Both Reid and Obama decided that a shutdown was in their political interests, because they knew that the media would echo their talking points, which blamed the Republicans. They also knew that a predictable number of Republican senators would parrot those same talking points.

Thus, every time the House voted to fund the government, Senate Democrats voted it down on a party-line vote, and the media dutifully repeated that it was *Republicans* who had shut down the government.

When the Democrats blocked funding, the shutdown began, but under law, only 17 percent of government spending (that deemed "nonessential") actually shut down. The remainder, the roughly 83 percent of government spending that is deemed "essential" (including Social Security and Medicare), continued uninterrupted.

The nightmare for Democrats was that they would "shut down" the government and nobody would notice. So they went out of their way to make the shutdown as painful for Americans as possible.

We all remember the absurd games the Obama administration played with visitors to the World War II Memorial in the nation's capital, instructing park rangers to erect barricades to try to stop veterans from visiting a monument to their heroism. I was proud to join a number of members of Congress welcoming these veterans to the monument during the shutdown. More than one of those World War II vets came carrying wire cutters. One said to me, with a twinkle in his eye, "We stormed the beaches of Normandy. Do they really think a couple of metal fences can keep us out?"

Once the shutdown began, Mike and I spent a lot of time trying to strategize how to win the fight—even though our Senate leadership was actively working against us. We studied what had happened during the last significant shutdown, in 1995, when Republicans had also been blamed.

Republican House members who had gone through the 1995 shutdown told us that some of the strongest pressure on them to surrender to President Clinton came from constituents upset about the closure of museums and parks. Families would come to Washington with their kids looking for a wonderful vacation and discover the Air and Space Museum wasn't open. They were furious.

Accordingly, we developed a strategy that tried to avoid those pitfalls with miniature continuing resolutions (CRs), or short-term bills that funded specific government programs. One mini-CR would fund national parks. Others would fund military salaries and benefits, the Department of Veterans Affairs, and the National Institutes of Health. Even if Democrats continued to force a shutdown, Republicans could use these mini-CRs to fund elements of the government that should not be held hostage to the Democrats' shutdown; we would dare the Democrats to vote against them.

Throughout the shutdown, Mike and I met repeatedly with key House Republicans to share strategy. The most notorious of those meetings occurred at Tortilla Coast, a popular Capitol Hill restaurant. Since we had chosen to meet in a public place a few blocks from the Capitol, reporters naturally saw us, and speculation swirled as to what on earth we could have been discussing. That Tortilla Coast meeting has entered D.C. lore, and it's a sign of how dysfunctional Washington is that it was deemed newsworthy that House and Senate members were actually meeting to work together.

At those early meetings, Mike and I pitched our "mini-CR"

strategy to House Republicans. They liked it, took it, and ran with it. And at first, it worked.

The first mini-CR the House passed was to fund the salaries for our military. For weeks, Harry Reid had been threatening that a shutdown would halt salaries for our soldiers. But when the House mini-CR came over, Reid blinked and allowed it to pass. The political risks were just too great for him to stop it.

After that, the House passed mini-CRs for a number of other priorities, including our national parks, for the VA, and for the NIH. Each of these, Reid and the Democrats killed.

If we had actually been trying to win, Senate Republicans would have mounted a concerted political campaign focused on these mini-CRs. For example, Senate Democrats had specifically objected to funding the VA. The CR for the VA said nothing about Obamacare; it just provided funding for the agency. Imagine if the Karl Roves of the world began running major ad buys in red states, like the following:

"Our veterans have risked it all for us, and yet Senate Democrats are holding the Department of Veterans Affairs hostage to try to force Obamacare on us. The House has passed legislation fully funding for the VA, but Democrats are blocking it. Call Senator Pryor (or Warner, or Begich) and tell him to stop playing games and fund the VA."

That's how we actually win fights, and it's how we would have put enough pressure on red-state Democrats to start to get them to flip. But Republican leadership wasn't interested in that.

Similarly, the House passed another mini-CR funding the NIH. And Democrats cynically blocked it. The vulnerability of their position was illustrated in the following press exchange between Harry Reid and CNN's Dana Bash:

"You all talked about children with cancer unable to go to clinical trials," Bash told Reid. "The House is presumably going to pass a bill that funds at least the NIH. Given what you said, will you, at least, pass that? And if not, aren't you playing the same political games that Republicans are?"

But Senator Reid didn't want to hear it, because he didn't really care about NIH's funding; he cared about preserving Obamacare at all costs. "What right do they have to pick and choose what part of government's going to be funded?" he responded. (The answer to that particular question is found in Article I, section 9, clause 7 of the U.S. Constitution, which provides, "No Money shall be drawn from the Treasury, but in Consequence of Appropriations made by Law. . . .")

To her credit, Bash did not meekly accept Reid's dismissal of both her question and Congress's power of the purse. "*But if you can help one child who has cancer,*" asked Bash, "*why wouldn't you do it?*"

"Listen," replied Reid in an irritated tone. "Why, why, why would we want to do that?"

It was a chilling answer, especially from someone purporting to be concerned about improving health care. He didn't help his cause when he condescendingly added, "To have someone of your intelligence suggest such a thing maybe means you're irresponsible and reckless."

If Washington Republicans had actually been playing to win, they would have welcomed the national debate with Reid about why he was going to hold hostage a child with cancer in order to force Obamacare on the American people. They would have aired Reid's statement in television commercials around the country. But they didn't do it, because Republican leadership didn't want to win.

Some readers might think that sounds a little harsh. You might assume they wanted to win, but just couldn't. Their behavior suggested otherwise.

Leadership showed their true intentions about midway through the shutdown. We were getting pounded on television (including by about twenty Senate Republicans) and leadership was doing nothing to put the heat on the Democrats. Nor did they do any polling to even try to see how we could win. So I decided to spend my own political money on national polling. The results were unsurprising: Republicans largely blamed Democrats for the shutdown, and Democrats blamed Republicans.

But, critically, Independents were nearly evenly divided (with just 6 points differential). And fully 22 percent of Independents were undecided. We tested multiple messages, and the polling showed that undecided Independents (and Democrats) swung sharply our way if they heard the following: "Republicans are working hard to fund government priorities, but Democrats are refusing to compromise or negotiate."

When the Senate Republican leadership grudgingly gave me a few minutes at our caucus's lunch to review the results of the poll, I explained that the numbers showed this was a strong winning message. Even if they were unhappy to be where we were, I urged, it was in all of our interests to win this battle.

The very next day, at a strategy session among the Senate Republicans, our leadership stood up and instructed Republicans to go on television with the following message: Harry Reid and the Senate Democrats are negotiating with us to compromise and end the shutdown.

Not only was this messaging not productive; it was the exact same message that Reid would have asked us to carry if he had looked at our polling.

———

At every stage in the battle to defund Obamacare I said that a compromise was possible, and a compromise would be anything that provided meaningful relief to the millions of men and women suffering under that most misguided of social experiments. Several weeks after the filibuster, I hoped President Obama would begin to discuss a solution when he invited Senate Republicans to the White House to discuss the shutdown. But I wasn't optimistic. This was not a president who plays well with others, including his own party in Congress, most of whom he ignores. It seemed unlikely that his invitation to the White House was an invitation to find common ground.

Nevertheless, I trooped dutifully up Pennsylvania Avenue and sat respectfully in the East Room of the White House with my Republican colleagues. President Obama entered the room and began one of the most bizarre meetings I've ever been in. He announced: I just want you to know I will not negotiate. I will not compromise. I will not give in on anything. The government will remain shut down until you agree to everything I want.

I've never seen anyone ask someone else to come over, only to explain that he's not willing to discuss, negotiate, or compromise on anything. But then again, we never have seen a president like Barack Obama.

The president was clearly on the same page with Harry Reid. He knew his media lapdogs would report that it was Republicans who were refusing to compromise. He knew the Democrats in Congress cared nothing about keeping the shutdown going as long as possible. And he knew the Republican leadership in the Senate would give him everything he wanted.

That's exactly what happened. On October 16, 2013, the Republican Senate leadership agreed to fund Obamacare in exchange for . . . nothing. Later that night, both houses of Congress voted for this surrender, which did nothing to protect the American people from Obamacare. The next day, President Obama signed it into law.

Fifty-two years after concerned Americans at coffee parties had listened to Ronald Reagan warn against socialized medicine, the radical left was now one step closer to its Holy Grail of a single-payer, government-run socialized health-care system.

———

The Obamacare defund fight could have turned out differently. Imagine, if you will, the entire fight with one minor detail changed: Imagine what would have happened if Republican Senate leadership had simply decided . . . let's support House Republicans. We may not like it, but let's stand together as Republicans, against the Democrats and against Obamacare.

With that one small change, we would have been able to focus at the outset on red-state Democrats. We would have been able to target ads, focusing on their blocking funding for veterans and cancer funding and a host of other vital priorities. We could have worked to energize the grassroots in their states to light up the phones to those Democrats.

And, if we had done so, there's a good chance quite a few of them would have gone to Harry Reid and said, "Get us out of this. We're getting killed back home. Unless you want to lose the majority, let's compromise."

Admittedly, unless it had gone perfectly, it probably wouldn't have stopped Obamacare altogether. But the odds are significant that we would have been able to achieve real, meaningful relief for the millions who have been hurt by this disastrous law.

But that didn't happen.

Even so, the battle accomplished a great deal. At the time, nearly every pundit in D.C. said that the shutdown would doom Republicans, that the Obamacare fight would ensure that Harry Reid remained majority leader.

They were wrong. Even with leadership undercutting us, the defund fight did something remarkable: It elevated the national debate over Obamacare. It highlighted the millions of Americans who were losing their jobs, being forced into part-time work, losing their health insurance, losing their doctors, and facing skyrocketing premiums. And it made clear that it wasn't because of nefarious insurance companies (the Obama administration's original culprit, upon whom they intended to blame the cancellations); it was because of this failed law.

In November 2014, Republicans won a historic victory at the polls. We won more seats in the House than any year since the 1920s. We won nine new Senate seats, and retired Harry Reid as majority leader. And the number-one issue that Republican candidates campaigned upon—the top subject of campaign ads across the country—was Obamacare.

Did it occur to anyone in Washington that our winning a tsunami of an election focused on Obamacare maybe, just maybe, was a direct result of our having energized and mobilized millions of Americans against it? Not for a moment.

But it's true nonetheless. And, even more important, I believe we've laid the predicate for ultimately repealing this law. As a result of the amazing efforts of the grassroots, of our working so hard to highlight the immense harms, support for Obamacare has plummeted. In November 2014, Gallup found that just 37 percent of Americans approved of Obamacare.

In March 2014, Pew found that 53 percent of Hispanics disapproved of Obamacare. And in December, a Fox poll showed that

57 percent of women disapproved of Obamacare, and 60 percent of young people (under 35) likewise disapproved.

This is going to be a central issue—if not the central issue—in the next election. And as a result of the tremendous mobilization of the grassroots—and our winning the substantive argument that the law is failing badly—I am convinced that, come 2017, a Republican president will sign legislation repealing every word of it.

We need real health care reform. But it should expand competition and empower patients, and disempower government bureaucrats from getting between us and our doctors. We should allow people to purchase insurance across state lines (which is currently illegal), which will in turn create a fifty-state national marketplace for low-cost catastrophic coverage. If you want more coverage, you want more choices and lower costs. Obamacare gives us fewer choices and higher costs.

We should expand health savings accounts, so we can save in a tax-advantaged manner for routine healthcare and prevention. And we should make health insurance portable, so it goes with you from job to job, which goes a long way to eliminating the problem of pre-existing conditions. High-risk pools at the state level can solve the rest of that problem.

Personal, portable, and affordable. That should be where we go after repealing Obamacare.

But if it's up to Washington, to the career politicians in both parties, that will never happen. That's why changing Washington—empowering the grassroots to demand something different—is so incredibly important.

CHAPTER 10

★

Obama's Vacuum of Leadership

Only 250 miles separate Cairo and Jerusalem—roughly the distance between Washington D.C. and New York City—and yet in the late autumn of 1977, the two cities couldn't have been farther apart.

When the Egyptian president's plane touched down for a historic visit to Ben-Gurion Airport, Egypt and Israel had been at war for decades. The last conflict, the Yom Kippur War, mushroomed into a regional conflagration that killed ten thousand and left many more casualties. By the time the guns went silent, Israel occupied Egyptian territory in the Sinai Peninsula, and a state of war endured. Only the slightest provocation on either side might invite more bloodshed and more death.

Which was what made the visit—after the sun had set and Sabbath had ended on Saturday, November 19, 1977—so stunning to the rest of the world.

When his feet touched the tarmac, Anwar Sadat, a trim, serious man with a black mustache and a receding hairline, had become the first Arab leader ever to visit the Jewish state in its entire existence. The dramatic pilgrimage was the culmination of months of secret negotiations.

The following day, President Anwar Sadat prayed at the al-Aqsa Mosque on the Temple Mount, visited the Church of the Holy Sepulchre and the national Holocaust museum, called Yad Vashem. Then he turned to the real business of the visit, a historic address to the Israeli parliament that only weeks ago seemed unimaginable.

"I come to you today on solid ground to shape a new life and to establish peace," Sadat said. "But to be absolutely frank with you, I took this decision after long thought, knowing that it constitutes a great risk."

It was a historic overture to peace, and what began that day in the Knesset would be consummated at the White House in 1979 with a historic peace treaty between Israel and Egypt. Sadat would also turn his country away from the Soviet Union and toward an enduring alliance with the West. Forging ties with the Jewish state and the West would earn Sadat enduring hatred in the Arab world, especially among Islamist radicals committed to Israel's destruction. Back in Cairo, the Muslim Brotherhood and its affiliated Islamist groups marched against Sadat's supposed betrayal.

On October 6, 1981, a radicalized Muslim officer broke formation in a presidential military parade in Cairo, reached for grenades from under his helmet, and threw them at Sadat. The Egyptian president died from his wounds several hours later.

Sadat risked his life, and sadly lost it, because he had the courage to speak the truth about the need for peace with Israel. He stood against the radical Islamists for whom nothing less than the destruction of the Jewish state was satisfactory. But his act of vision and courage still echoes through the Middle East.

————

On January 21, 2009, Hillary Rodham Clinton was sworn in as the nation's sixty-seventh secretary of state.* Standing alongside President Obama, the new secretary vowed to bring American "smart power" to bear on the problems of the world. Four years later, she left office with America's international standing in tatters. In the years since, the situation has deteriorated even further.

Our world is far more dangerous now than it was when President Obama took office. His Nobel Peace Prize notwithstanding, peace is receding today faster than it has in a generation. President Obama and Secretary Clinton projected weakness, and weakness has proven provocative. Today, Russian president Vladimir Putin is on the march in Ukraine and eyeing the Baltic states. China is making an aggressive effort to exert global power by intimidating U.S. allies and demanding new territorial concessions, from South Korea to Japan to the Philippines to Taiwan and Singapore. Cuba is exporting arms to North Korea.

Consider the example of Egypt, which in the years since Sadat has grappled with violence, chaos, and extremism. Across the globe, there are hundreds of millions of peaceful Muslims, but a committed minority are using radical Islam to promote murder, torture, and barbarity. In recent months, there was a leader who courageously took rhetorical aim at the radicals corrupting the Islamic faith and posing a danger to the free world. That leader, sadly, was not an American president.

Though it barely registered in Western newspapers, on the first day of 2015, General Abdul Fattah al-Sisi spoke out about the need

—————————

*In keeping with the clubby pattern of the Senate, Secretary Clinton was confirmed by a broad bipartisan vote of 94–2.

to confront the radicals within the Islamic faith. The Egyptian president, himself a Muslim, called for "a religious revolution." He challenged his fellow Islamic leaders to stand against the radicals who use their faith to kill innocents. "It's inconceivable that the thinking that we hold most sacred should cause Muslims worldwide to be a source of anxiety, danger, killing and destruction for the rest of the world," he said. While al-Sisi may not be a champion of democracy, his initiative and courage is keeping the Muslim Brotherhood at bay. And, like Sadat, he is bravely risking his life to speak the truth.

Ironically, five years earlier, President Obama also spoke in Cairo—and at the very same institution. But his message was nearly the reverse. His remarks, called "A New Beginning," were widely seen in the Arab world as aligning the United States with the Muslim Brotherhood.[1] In effect, he apologized for America's role in the region before his presidency and seemed to share the grievances of the anti-American radicals who, only months later, would topple the Egyptian government.

In the months and years of chaos that followed the so-called Arab Spring, the terrorist group ISIS reclaimed much of Iraq, crucifying Christians, beheading journalists, and imposing a reign of terror. And Iran continued its relentless pursuit of nuclear weapons.

There was also the matter of four dead Americans at the U.S. compound in Benghazi, Libya: Ambassador Chris Stevens, the first U.S. ambassador killed on duty since the Carter years; foreign service officer Sean Smith; and retired Navy SEALs Tyrone Woods and Glen Doherty. The September 11, 2012, attack on the Benghazi compound was coordinated and carried out by radical Islamic terrorists. Secretary of Defense Leon Panetta testified to the Senate that he knew "immediately" that it was a terrorist attack. And yet for weeks President Obama and Secretary Clinton insisted instead that it was a spontaneous protest over an Internet video.

The administration's feckless response to Benghazi was emblematic of President Obama's long-standing approach to radical Islamic terrorism—three words that almost never enter his vocabulary in the same sentence. In his worldview, the real root problem behind terrorism is disaffected youth who have been antagonized by American and Western imperialism. He and his administration dogmatically refuse to call terrorism "Islamic" or "Islamist," nor will they reference "jihad."

Thus, when a jihadist gunned down killed down fourteen people at Fort Hood, Texas, the administration called it "workplace violence." When Islamists murdered Jews in a kosher deli in Paris, Obama described their anti-Semitic rampage as simply "random" violence. When ISIS beheaded twenty-one Coptic Christians, the White House suggested it was merely because they were "Egyptian citizens." To the contrary, they were murdered because of their faith, and, as Pope Francis powerfully observed, "their blood confesses Christ."

It is impossible to defeat an enemy when you can't even admit that it exists.

———

Of all the national security threats facing America, none is greater than the continued efforts of Iran to acquire nuclear weapons capability. And yet, instead of acting resolutely to ensure that can never happen, President Obama has relaxed sanctions and entered extended negotiations that have dramatically increased the chances of a nuclear Iran. The president's deputy national security advisor, Ben Rhodes, has described an Iranian nuclear deal as "probably the biggest thing President Obama will do in his second term on foreign policy." He compared it to Obamacare, for the second term. (I think he meant that as a compliment.)

Toward that end, Obama has encouraged a "new beginning"

with Iran, where since 1979 Shiite mullahs have forged one of the world's most brutal, anti-American tyrannies. Indeed, Iran has for thirty years positioned itself as the implacable foe of America and our allies. Every year on "Death to America Day"—yes, that's an actual day—the Iranian government celebrates the anniversary of the hostage-taking at the U.S. embassy in Tehran. Iran has also functioned as a leading state sponsor of terrorism, overseeing deadly attacks from Beirut to Buenos Aires. And even under severe economic sanctions that crippled the economy, the mullahs have been pursuing both nuclear weapons and the short- and long-range missiles to deliver them against such sworn enemies as Israel, which the Iranians have vowed to erase from the earth.

But the president was convinced he could reverse three decades of Iranian hostility with conciliation, understanding, and a perceived indifference to Iranian atrocities. This is a naïve worldview—the same liberal credulity espoused by Jimmy Carter and others, until they are awakened to the folly of coddling and excusing extremism. Sadly, President Obama hasn't learned that lesson. When, in 2009 in response to the fraudulent reelection of Mahmoud Ahmadinejad, Iranians poured into the streets to demand basic democratic freedoms and the mullahs brutally crushed the revolt, there was scarcely a word from the United States. And the episode certainly didn't change Obama's approach. Over the next few years he tried to forge a friendship with Iranian radicals, sending personal letters to Ayatollah Khamenei and reaching out by cell phone to Iranian president Hasan Rouhani.

Eventually, in the fall of 2013, in response to Obama's persistence, the Iranians agreed to the first direct negotiations between our two governments since the extremists took power. The agreement, the administration claimed, was a "first step" toward addressing "our concerns over the Islamic Republic of Iran's nuclear program."

Of course, this dialogue with Iran was something Obama had

pledged to undertake since he first ran for the White House. But one cannot negotiate from weakness—especially with bullies and tyrants.

America has always believed in peace through strength. It was a mistake, a profound mistake, to lead the world in relaxing sanctions against Iran. That was at the outset of the negotiations, before anything real had been agreed to. As a result, billions of dollars have already flowed into Iran. All the while, Iran continues building centrifuges, enriching uranium, and developing its ICBM program.

In the 1990s, the Clinton administration followed the same pattern, leading the world in relaxing sanctions against North Korea; as a result, billions of dollars flowed into that country, and they used that money to develop nuclear weapons. Ironically, the Obama administration recruited the very same person—Wendy Sherman—who had led the failed North Korea talks to become our lead negotiator with Iran. But here the results are likely to be far worse.

With North Korea, both Kim Jong Il and Kim Jong Un were radical and unpredictable, but both father and son were and are megalomaniacal narcissists. That means some degree of rational deterrence is possible, because neither was willing to risk losing power. Supreme Leader Khamenei and the Iranian mullahs are religious radicals, who embrace death and martyrdom. As a result, ordinary cost-benefit analysis is far less reliable.

Here are some of the statements Khamenei has made about Israel:

- "Israel is the sinister, unclean rabid dog of the region."
- "The foundation of the Islamic regime is opposition to Israel and the perpetual subject of Iran is the elimination of Israel from the region."
- "Israel's leaders sometimes threaten Iran, but they know that if they do a damn thing, the Islamic Republic will raze Tel Aviv and Haifa to the ground."

Hassan Tehrani Moghadam, who has been described as the brains behind Iran's ballistic missile program (he was assassinated in 2011), said the following in his last will and testament: "Write on my tombstone: This is the grave of someone who wanted to annihilate Israel."

Given this religious zealotry, if Iran were to acquire nuclear weapons capability, there is a real chance its leaders would actually use those weapons, either in the skies of Tel Aviv or even on New York or Los Angeles.

Even if Iran ended up not using its nuclear weapons, the immediate result would be that the Arab countries of the Persian Gulf would move to acquire their own. We would see nuclear proliferation across the most dangerous region of the world, with governments that often have shady ties to radical Islamic terrorist groups. It would be a powder keg, waiting to explode. And that is the *best-case scenario*.

And, it is worth noting, the Iranian mullahs' hatred is not just limited to Israel. At a large public rally in 2014 (right in the middle of the Obama negotiations), Khamenei cried out that the United States is "the greatest violator of human rights in the world." The crowd responded, chanting in unison, "Death to America!"

One of the simplest principles of geopolitics: When somebody tells you they want to kill you, believe them. And yet Obama and Clinton and Kerry persist in believing that we can reach a reasonable common ground with these Islamic hard-liners.

If we are to negotiate, fine; but we need less carrot and more stick. For that reason, I've introduced legislation to immediately reimpose sanctions, and to strengthen them even further. That legislation lays out a clear path to lifting the sanctions: Iran must 1) disassemble all 19,000 centrifuges; 2) hand over all of its enriched uranium; 3) shut down its ICBM program (which exists for the sole purpose of striking America); and 4) cease being the world's leading state sponsor of terrorism.

The Obama negotiations presuppose a "right to enrich." Individuals and nations have long been recognized to have rights to life and liberty, but there is no natural right to enrich uranium. And Iran's conduct had demonstrated it cannot be trusted with the materiel for nuclear weaponry.

Moreover, any relaxation of sanctions should occur only *after* Iran releases the American citizens it has unjustly imprisoned. For example, Iran has sentenced American pastor Saeed Abedini to eight years in prison, for the "crime" of building an orphanage and peacefully practicing his Christian faith. Convicted of attempting to undermine the Islamic state, he has endured beatings in prison and the denial of medical care. Throughout his ordeal, Pastor Saeed has led dozens of his captors and fellow prisoners to Christ.

And while American negotiators met and broke bread with Iranians in Paris and Geneva and Vienna, Iran chose to transfer Pastor Saeed from the brutal Evin prison to the even more horrific Rajai Shahr, where they house their death row. The day of the transfer? "Death to America Day."

We should demand the release of Pastor Saeed, along with his fellow prisoner of conscience Amir Hekmati, imprisoned journalist Jason Rezaian, and yet another American, Robert Levinson. Meanwhile the Obama administration focuses more on a misguided nuclear deal than on protecting our nation or freeing American citizens from wrongful captivity.

We have two more years of President Obama's presidency, and the foreign policy wreckage across the globe increases every day. My approach in the Senate has been to try to do everything possible to minimize the damage.

I've worked hard to try to build bipartisan agreement, addressing some of the clearest challenges to national security. And, by focusing on bipartisan approaches to national security, we've had remarkable success.

For example, last year, Iran named Hamid Aboutalebi to be its ambassador to the United Nations. Aboutalebi is a known terrorist who participated in holding Americans hostage in Tehran for 444 days in 1979 and 1980 (he claims he was "just a translator"). His nomination was a deliberate slap in the face to the United States by the Islamic Republic.

In Washington, there was much consternation over Iran's belligerent actions. Yet people on both sides of the aisle were resigned that there was nothing we could do about it. International norms dictated we must accept any ambassador that a member nation of the United Nations has selected. While there was precedent for blocking particularly odious foreign leaders from entering the United States for UN General Assembly meetings, ambassadors had not been blocked. It just wasn't done.

It seemed to me that, regardless of protocol, it was altogether unacceptable to have one of the 1979 Iranian hostage takers walking the streets of Manhattan with diplomatic immunity. It was absurd to see Iran exploiting the UN to insult America—while, I might add, we were engaged in direct negotiations over their nuclear program.

I wanted to find a way to stop Aboutalebi from being admitted. Our research uncovered an existing statute that allowed the Department of State to reject a visa application for an ambassador on the grounds that he had been a spy. We drafted legislation amending that statute to also include former terrorists.

Our legislation received support from senators as varied as Lindsey Graham, the South Carolina Republican, and New York

Democrat Chuck Schumer. Indeed, when Schumer sang my praises on the Senate floor, I came up to him and said with a laugh, "Chuck, you better be careful. . . . Lightning's going to strike you!"

Our legislation passed the U.S. Senate 100 to nothing. It then went to the House of Representatives, where it also passed unanimously. Then, on the afternoon of Good Friday, President Obama signed it into law.

As a result, a known terrorist was not allowed into the United States, and in January 2015 Iran named another nominee to be its UN ambassador.

A couple of weeks later, the president spoke at the annual White House Correspondents' Association dinner. This was one of those elite media events where the president is expected to give a humorous speech in front of the black-tied glitterati. In the speech, Obama observed, "Just a couple of weeks ago Ted Cruz introduced legislation that I signed into law. Here's a picture of the signing ceremony." He put up a picture of himself, of me, and of the Devil . . . and Hell freezing over.

I've also worked hard to protect U.S. sovereignty, a deep passion of mine ever since we prevailed over the UN and the World Court in *Medellín v. Texas*. Toward that end, I was very glad to be able to significantly modify legislation to prevent the diminishment of America's authority at the International Monetary Fund. Specifically, when Russia invaded Ukraine, the world rightly decried that as an act of war. Congress responded by imposing sanctions on Russia and also offering an aid package to our friends in Ukraine. But then the Obama administration cynically decided to use that Ukrainian aid package as an excuse to force through a long-standing and irrelevant priority: the administration's expansion of the IMF.

The proposed "reform" would have increased the exposure of U.S. taxpayers by billions of dollars. It would have undermined the influence of the United States at the IMF and decreased our voting share. And, perversely, it would have *increased* the influence in voting share of Russia. That was certainly a strange way to punish Vladimir Putin for invading Ukraine.

When I went to the Senate floor and objected to the IMF reforms, my objection was met with scorn. Senators from both parties denounced me for slowing down the Russia sanctions bill. The passionate and at times volcanic John McCain—with whom in the past year I've actually struck up an unlikely friendship—excoriated me on the floor. His most reliable ally, Lindsey Graham, pulled him aside and pointed out humorously, "John, it's not like Ted is actually driving a Russian tank into Ukraine!"

My objection stopped the legislation that day, and over the ensuing week more and more public attention was focused on the problematic IMF reforms. I worked with Rand Paul on a letter to Harry Reid arguing that these provisions were preventing us from getting strong bipartisan legislation out of Congress on Ukraine. We urged our friends in the House not to go along with giving up U.S. sovereignty to an unelected international body. And it worked. Just one week later, President Obama and Harry Reid surrendered entirely, stripping the IMF provisions from the legislation. The bill then passed the Senate, 98–2.* Without my objection, the misguided IMF "reform" would have likely passed as well. What passed instead was far better for U.S. interests.

I was also able to introduce and pass into law several positive amendments to the National Defense Authorization Act, in 2013 and again in 2014. In particular, I teamed with Mike Lee to increase the protection of religious liberty in the military, and to protect

---

*Ultimately Senator Paul decided not to vote for the Russia sanction legislation, but I appreciated his help in strengthening it.

chaplains from threats to their faith; mandated that the Defense Department address the growing threat of missile attack from the Gulf of Mexico; joined with John Cornyn to prohibit the military from purchasing Russian-made helicopters; invited Taiwan to join in the Navy's Rim of the Pacific military exercise; and prohibited a new domestic base realignment and closure (BRAC) process until the Defense Department first conducts an overseas BRAC process.

I was also proud to introduce and pass an amendment to allow the victims of Nidal Hasan's terrorist attack on Fort Hood to receive the Purple Heart. Major Hasan murdered fourteen innocent souls (including an unborn child), while shouting "Allahu Akbar." He had earlier communicated with a radical Islamic cleric, Anwar al-Awlaki, to inquire about the permissibility of jihad against his fellow soldiers. And yet, inexplicably, the Obama administration refused to call the attack terrorism, instead classifying it as "workplace violence." For five years, the victims of the Fort Hood attack were wrongly denied the Purple Heart. But in 2014, over the strenuous objections of the administration, I was able to win bipartisan support for my amendment. And, in 2015, the Army finally awarded the Fort Hood victims the Purple Heart.

Another deep and abiding passion has been strengthening our friendship with the nation of Israel, an alliance that has been profoundly undermined in the Obama administration. Unfortunately, over the past six years, the Obama administration has demonstrated an unprecedented hostility to the Jewish state, and its actions have weakened not only our alliance but Israel's very security.

I believe we have no greater friend in the Middle East. Like us, Israel is a nation of immigrants, a country based on ideas, on our shared Judeo-Christian, democratic values. I also believe that from a purely American point of view, supporting Israel is tremendously beneficial to our national security interests. It is also in our economic interests. Since being elected, I have visited Israel three times and have worked hard to repair the damage the administration has done to this vital relationship.

In the summer of 2014, three boys—Naftali Fraenkel, Gilad Shaar, and Eyal Yifrach—were kidnapped by Hamas terrorists in Israel. These Jewish teenagers went missing for three agonizing weeks. Naftali Fraenkel had family in Brooklyn. I had the opportunity to sit down with his aunt, a lovely woman who was praying fervently for her nephew.

Fourteen days into their abduction, I gave a Senate speech alongside photos of each of the three boys. As Stalin chillingly observed, one death is a tragedy, and a million a statistic. My hope was to personalize these boys, to make them real. Naftali, age sixteen, played guitar, and enjoyed Ping-Pong. Gilad, also sixteen, enjoyed scuba diving and cooking, and liked to bake pastries for his five sisters. Eyal, the oldest at nineteen, loved sports and singing. I called on the Hamas terrorists who had taken them—whose leaders were publicly celebrating—to give these boys back.

I did not know it at the time, but the boys had already been murdered: shot, execution-style, by Hamas.

In response, I introduced legislation providing for the State Department to offer a $5 million reward for information leading to the capture of the terrorists who kidnapped and murdered Naftali Fraenkel. Because Naftali was a dual U.S.-Israeli citizen, there was precedent under the Rewards for Justice program. That legislation was cosponsored by New Jersey Democrat Bob Menendez, then the chairman of the Senate Foreign Relations Committee. The Cruz-Menendez legislation passed the Senate unanimously.

Companion legislation was introduced in the House, and it was on a path to passage when, thankfully, the Israeli military captured the Hamas terrorists responsible.

After the boys' bodies were discovered, the world saw rockets rain down upon Israel. The Israel Defense Forces discovered terror tunnels dug underneath the border from the Gaza Strip, coming up into Israeli schools and other civilian targets. Hamas was planning a massive attack—apparently on the holy day of Rosh Hashanah—in which thousands of terrorists would pour into Israel to kidnap children and murder Jews while they celebrated the holiday.

Israel acted as any sovereign nation would in self-defense. The IDF moved into Gaza and attacked Hamas strongholds and destroyed the tunnels. Yet doing so was rendered significantly more difficult by the international outcry over the inevitable civilian casualties. The IDF took extraordinary measures to minimize such casualties, going so far as to drop leaflets warning civilians to evacuate specific targets. But Hamas managed to dramatically increase the number of casualties by preventing civilians from fleeing and by deliberately using women and children as human shields. They even located the Hamas headquarters in the basement of a hospital. Hamas also placed their rockets in UN schools.

I thought this was a key example of how little the terrorist group cares about the well-being or even the lives of the Palestinian people. So I joined with New York Democrat Kirsten Gillibrand in introducing a resolution condemning Hamas's use of human shields as a war crime. Our resolution passed the Senate unanimously, and then passed the House unanimously as well.

Another example of successfully changing administration policy concerned the Federal Aviation Administration's ban on U.S. flights to Israel. At the height of the rocket attacks, the administration did something unprecedented. After a Hamas rocket—one of the very few not intercepted by the marvelous Iron Dome missile defense system—fell about a mile from Ben-Gurion Airport, the FAA announced a ban on U.S. airlines flying into the country. During the entire history of Israel, even in times of war and turmoil, the FAA had never before banned such flights. Such a ban was a serious blow to Israel's economy at a moment when the country was under attack.

So I asked a very simple question: Has the Obama administration launched an economic boycott on Israel? The FAA, I pointed out, did not ban flights to Pakistan or Yemen or Afghanistan. The FAA did not even ban flights into much of Ukraine, despite the fact that only a month earlier a passenger airplane had been shot down by a Russian BUK missile. Why was Israel being singled out?

And why was it precisely timed to coincide with John Kerry's arriving in the Middle East with $47 million in aid for Gaza, aid that would inevitably end up in the hands of Hamas? Why was the administration putting economic pressure on Israel while at the same time using political pressure to stop their military campaign of destroying the rockets and tunnels?

Within an hour, a State Department spokeswoman was asked the question I had posed: whether the FAA's ban was in fact an economic boycott of the nation of Israel. She dismissed the suggestion as "ridiculous" and refused to address questions about the FAA's unprecedented actions. I therefore announced that I would place a hold on all State Department nominees until they answered these pressing questions.

In the interim, the roar of criticism of the administration's punitive air travel ban grew louder. New York's then-mayor Michael Bloomberg, with whom I disagree on many other issues, took the bold step of traveling to London to take a commercial flight to Tel Aviv, in order to demonstrate how safe it was.

And, the heat and light was so intense, within thirty-six hours, the FAA had no choice but to publicly lift the ban and allow flights to resume.

Standing for Israel is important, and it is sometimes controversial.

In the summer of 2014 my office was approached by a new group called "In Defense of Christians," the ostensible purpose of which was to bring together the various Christian sects of the region, who had a history of internal squabbling, so they could present a united front at a conference in Washington in September. The Syrian civil war had been raging for three years, and as the civil war dragged on, jihadists from Al Qaeda, Jabhat al-Nusra, and a new group calling itself the Islamic State in Iraq and Syria (ISIS) attacked Christians with impunity, torturing and murdering them in a systematic campaign of genocide. Our office made it be known that we were interested in helping any way we could. "IDC" invited me to keynote their gala dinner, and because a number of people I hold in the highest respect including my undergraduate thesis advisor Professor Robby George of Princeton were participating, we agreed.

The event seemed straightforward until the day before the dinner, when news stories broke that suggested the funding for the conference had ties to Hezbollah—and so Iran—and that participants had been requested not to talk about Israel or persecution of the Jews in their remarks. As controversial as the event might be, cancelling just didn't sit right with me. For starters, the issue was a critical one, and there weren't enough loud voices speaking up for these innocent Christians who were being crucified and beheaded by Islamic radicals. In addition, I had said I would go and I hated going back on my word. I didn't want to cede the discussion to the radicals. I decided I would go, give my speech, but speak the whole truth.

I gathered my thoughts as we drove up to the Omni Shoreham Hotel. My prepared remarks emphasized the importance of standing for all persecuted religious minorities in the Middle East—Christians, Jews and moderate Muslims—and not making a deal with devils such as Hamas, Hezbollah or Assad who cannot be trusted. This line of reasoning was fine as far as it went, but the minute I stepped into that banquet room I knew it wasn't going to fly. There was an unsettled feel, and a lot of security. We were taken around the side to wait backstage.

When I stepped onstage, at first the welcome was warm. But as I began my remarks, and made a mention of the Jews, and then Israel, a muttering started that quickly became hissing and booing. To be fair the majority of the audience was either listening respectfully or even applauding, but the minority was much more vocal. So I told them Christians in the Middle East had had no greater friend than Israel, which made them even angrier.

I set my prepared remarks aside and spoke from the heart, "Those who hate Israel hate America. And those who hate Jews hate Christians." The boos grew louder. "If those in this room will not recognize that, then my heart weeps. If you hate the Jewish people, you are not reflecting the teachings of Christ. And the very same people who persecute and murder Christians right now, who crucify Christians, who behead children, are the very same people who target Jews for their faith, for the same reason."

The shouting was so loud the head of the conference came on stage to ask them to give me a fair hearing. They would not. So I simply said, "If you will not stand with Israel, then I will not stand with you. God bless you, good night," and walked off.

Surrounded by burly security guards and police, our party was escorted through the kitchen. When we got into the hotel lobby we were pursued by various attendees, some expressing support, others more aggressive outrage. The IDC incident hit the press shortly thereafter, with some hailing me as a hero, and others vilifying me as a self-promoting opportunist who had planned the whole thing.

I was in fact neither. As sad as I was that the purported message of the conference—to stand together in defense of Christians—had been undermined, I had no regrets. All I did was tell the truth. You can't beat hate with hate. And the Christians of the Middle East won't be helped if we turn our back on a true friend and try to make common cause with an enemy that, if it succeeds in destroying the Jews, will not stop there.

———

We can continue to work to mitigate the damage for the next two years. But ultimately, we need a fundamentally new direction for American foreign policy.

The Obama administration's efforts to forge "a new beginning" with Iran might well mean that one of the most determined enemies of America will possess a nuclear weapon by the end of Obama's term. Here's the grim assessment of Ali Younesi, senior advisor to President Rouhani and formerly Iran's intelligence minister:

> Obama is the weakest of U.S. presidents; he had humiliating defeats in the region. Under him the Islamic awakening happened. . . . Americans witnessed their greatest defeats in Obama's era: Terrorism expanded, the U.S. had huge defeats under Obama and that is why they want to compromise with Iran.

When President Obama leaves the White House, he will leave behind a legacy of ashes in international affairs. The vacuum of leadership created by Barack Obama and Hillary Clinton will take a generation to restore by reversing the effects of massive defense budget cuts and reinstilling confidence in our friends and allies that America foreign policy is once again guided by moral clarity.

It is not impossible, however. And in fact there is a parallel. When Jimmy Carter left the White House in 1981, America was in a similar state of paralysis, incoherence, and crisis. Iran was an ascendant power, similarly mocking the perception of American weakness. The Russians had invaded a neighbor—Afghanistan—and threatened others in their orbit. That all changed when a confident president, Ronald Reagan, willing to exert American leadership and defend American values, took the reins.

Likewise, I believe with true leadership—and a president unafraid to face down tyranny and call evil by its name—we can see the restoration of America to its rightful place as the leader of the free world.

———

As abysmal as Obama's record abroad has been, the damage he has done to our Constitution is just as devastating.

In the history of our republic, we've never seen a president so willing to ignore the law, to refuse to enforce the law, to defy the law, or to claim the authority to outright change the law as Barack Obama. As liberal law professor—and former Obama supporter—Jonathan Turley has said, "What's emerging is an imperial presidency, an über-presidency as I've called it, where the president can act unilaterally."[2]

President Obama seems uncomfortable with the rule of law as laid out in our Constitution. He fundamentally rejects two of its most important principles: that the executive is required to enforce all the laws, even those he dislikes; and that there are limits on the power of the federal government.

Our Constitution states that the chief executive may not negate congressional acts by picking and choosing which laws to enforce. Instead, it mandates that the president "shall take care that the laws be faithfully executed." The Founders understood that representative government is only representative if the executive enforces the laws passed by the people's representatives.

Our Constitution also enumerates the federal government's powers, which means that federal authority is limited to those relatively few powers listed by the Framers. The Tenth Amendment makes clear that all unenumerated powers belong "to the States . . . or to the people." And the other amendments in the Bill of Rights provide express protection against exercises of federal power that infringe on rights like the right to freedom of speech, the right to religious liberty, and the right to private property.

Although there are many elements of the Obama presidency that are troubling—the explosion of a crippling debt; the job-crushing effect of higher taxes and new regulations; the governmental takeover of our health care system—nothing about it has been more damaging than the president's repeated acts of lawlessness and his refusal to acknowledge long-respected limits on the power of the federal government.[3] It is a betrayal of our constitutional system.

As the ranking member of the Senate Judiciary Committee's Constitution Subcommittee, I released a series of reports chronicling seventy-six acts of lawlessness by the administration.

One of the most egregious examples of lawlessness is the president's nonenforcement of his signature legislative accomplishment: Obamacare. President Obama and the Democrats have repeated the mantra, "Obamacare is the law of the land."[4] Yet members of his administration have consistently been unwilling to abide by the plain text of the law they wrote with their Democratic allies in Congress.

For example, Obamacare's employer mandate requires employers with fifty or more full-time employees to pay a tax if they do not provide health insurance for their workers. Just as many individuals must obtain insurance, many businesses must provide insurance. This puts an unbearable burden on many entrepreneurs, disincentivizes job creation, and encourages employers to reduce the number of full-time employees.

The employer mandate was supposed to kick in on January 1, 2014, but in late 2013, President Obama decided to treat his buddies in big business differently than individual citizens subject to the individual mandate. He granted those businesses a one-year waiver from the employer mandate. Then, in February 2014, he extended many of the waivers through 2016—two years later than individuals were forced to comply with the law.

The president did the same thing for members of Congress. The text of Obamacare says that members of Congress will be on the exchanges, without subsidies, just like millions of Americans. Harry Reid and the congressional Democrats were horrified at that prospect, and they asked the president for relief. And so, contrary to law, the administration issued a ruling exempting members of Congress (but not ordinary Americans) from the explicit requirements of Obamacare.

Some people might wonder why an opponent of Obamacare like me is bothered by the president's failure to enforce a law I dislike. To be sure, I believe Obamacare is bad, and I'm fighting for Congress to repeal it altogether. But regardless of whether a law is good or bad, a president's unilateral revision of it is illegal—and dangerous to the rule of law.

The separation of powers requires the president to "take care" that *all* the laws passed by Congress are executed—the good, the bad, and the ugly. Obama's delaying decree is just as lawless as if a Republican president decreed a flat tax (good underlying policy) by unilateral edict.

Obamacare is far from the only law the president refuses to enforce. Two weeks after the 2014 midterm elections, in which President Obama's policies and party suffered a historic defeat and Republicans won both Houses of Congress, our forty-fourth president strode to a podium in the White House to make a stunning announcement: He would no longer enforce the immigration laws of the United States, and would in effect grant some four million illegal immigrants in the United States a blanket amnesty.

In 2011, President Obama had disclaimed any power to grant amnesty by executive fiat, telling the National Council of La Raza that the "idea of doing things on my own is very tempting . . . but that's not how our system works. That's not how our democracy functions."[5] As he put it, "I'm not the emperor of the United States."[6]

But a year later, in 2012, the president showed that his real understanding of "how our democracy functions" is not very democratic at all. He created an application process to grant amnesty to those who crossed the border illegally as children. This was simply an executive decree of legislation that Congress had already *rejected*.

The consequences of Obama's illegal amnesty became plain in 2013 and 2014, when we saw the numbers of unaccompanied minors crossing the border skyrocket. In 2011, roughly 6,000 unaccompanied children crossed the border. By 2014, that number had risen to 90,000. And by 2015 it is estimated to rise to 145,000. This was the predictable consequence: When you grant amnesty to those who come here illegally as children, you create an enormous incentive for more and more children to come illegally. And those kids are not being transported by well-meaning social workers. They are being trafficked by transnational criminal cartels. Subjecting tens of thousands of children to physical and sexual abuse, to victimization by vicious so-called coyotes, is the opposite of compassionate. And it is contrary to law.

And, in 2014, President Obama expanded amnesty even further, to an additional four million people, in direct defiance of the results at the polls in November 2014. As Obama rightly noted, his "policies [were] on the ballot" all across the country. The American people overwhelmingly voted against amnesty, and the president angrily decreed it anyway. The president's amnesty decree was lawless, and it was consistent with his politicizing immigration policy. I helped lead the fight against the "Gang of 8" amnesty bill because it would have made the problem worse. But the way to actually pass immigration reform is to focus on areas of bipartisan agreement: secure the border, now, and improve and streamline *legal* immigration. Indeed, there is no stronger advocate of legal immigration in the Senate than I am. But rule of law matters also.

The president also unilaterally repealed a key provision in the welfare-to-work reform that President Clinton signed into law in 1996. The premise of welfare reform in the 1990s was that those who benefit from our social safety net should be required to prepare for work, look for work, or work. But in 2012, the Obama administration announced that it would issue waivers of this work requirement. The waivers had no basis in law, but that did not matter. It was yet another illustration of President Obama's philosophy, which is best summed up in one of his favorite, repeated threats: "If Congress won't act, I will."[7]

As liberal law professor Turley has observed, "President Obama [has become] the president [that] Richard Nixon always wanted to be."

Fortunately, we still have at least some checks on the president's power. Indeed, the Supreme Court has unanimously rebuffed the president's vision of government power—not once, not twice, but more than *twenty separate times*;[8] however, the unprecedented volume of unanimous repudiations of the administration—by a Court that includes four liberals, two of them appointed by Obama himself—underscores what a radical vision of federal power the president is trying to assert.

A review of the aggressive assertions of government power made by the Obama administration is nothing short of breathtaking.

If President Obama had his way, the government could secretly put a GPS tracking device on the car of any American, without cause or suspicion of a crime. The FBI could literally set up a room straight out of a science fiction film, where a giant screen shows moving dots, and a hub of computers allows federal agents to know the location of every car of every family in the United States. If that sounds extreme, well, the administration defended precisely this interpretation of the Fourth Amendment's privacy guarantees before the Supreme Court in *United States v. Jones.* It lost unanimously.

If President Obama had his way, the police could not only follow your movements; they could also search through the smartphone of anyone arrested, even if there is no reason to suspect that the phone contains evidence of a crime. In other words, if you are carrying a smartphone when you are arrested for jaywalking, an agent of the state can read every email on your phone, look at every picture on it, and listen to every voice mail on it. (They can also check the numbers of every person you've called in your "recent calls"—although the NSA may already have that list.) The administration defended this extreme interpretation of the Fourth Amendment before the Supreme Court in *Riley v. California.* It lost unanimously.[9]

If President Obama had his way, Congress could pass laws that block the ability of houses of worship to choose their own ministers. If Congress wanted to force an Orthodox Jewish synagogue or a Catholic church to allow women to become rabbis or priests, no "ministerial exception" inherent in the Free Exercise of Religion Clause would protect religious institutions. The administration defended this radical interpretation of the First Amendment's freedom of religion in *Hosanna-Tabor v. EEOC.* It lost unanimously.

In *Hosanna-Tabor,* the Obama Justice Department argued that the First Amendment was irrelevant to the government's assertion of power over the church. Justice Elena Kagan—who was nominated to the Court by President Obama—remarked from the bench that it was "amazing" that the administration believed that "neither the

Free Exercise Clause nor the Establishment Clause has anything to say about a church's relationship with its own employees."

If President Obama had his way, the president could entirely evade the Constitution's requirement of senatorial "advice and consent" for nominations. The issue arose when Obama wanted to appoint nominees without sufficient support in the Senate for confirmation. Instead of finding new nominees with broader support, Obama claimed that the president has the power to unilaterally declare the Senate to be in recess—even if the Senate disagrees. Then, having declared a recess, Obama made appointments to the National Labor Relations Board and the Consumer Financial Protection Bureau by invoking his power to make temporary appointments during a Senate recess. Before the Supreme Court, the administration defended this unprecedented interpretation of the Recess Appointments Clause in *NLRB v. Noel Canning*. It lost unanimously.

Taken together, these cases illustrate a startling misunderstanding of privacy, religious freedom, federalism, and executive power. But perhaps no case exemplifies Barack Obama's unlimited vision of the government's power over its citizens better than the case of *Sackett v. EPA*.

The story of this obscure dispute began when the Sackett family wanted to build their dream home. They bought a vacant lot in Idaho. Some distance from their house site—beyond several other lots with buildings on them—was a lake.

After finding a drainage problem on their lot, the Sacketts filled part of their property with dirt and rocks and began to build their home. That's when they received a letter from the Environmental Protection Agency informing them that they had filled in statutorily protected "wetland." Without affording the Sacketts any opportunity to explain why their property was in no way a "wetland," the letter ordered them to stop construction, remove the fill, turn over a host of records, and allow the EPA access to the property. If the Sacketts so much as hesitated, they were subject to a $75,000 fine *for every day* they delayed.

The EPA's worst abuse of power was not its designation of the Sacketts' backyard as a wetland or even its threat of crippling fines. Unfortunately, that's business as usual for the modern EPA. But the case made it to the Supreme Court because the Obama administration went even further. It claimed the Sacketts should also be prohibited from taking the EPA to court to defend their property rights. All nine justices rejected President Obama's vision of faceless bureaucrats being able to take away Americans' constitutional right to property without so much as a judicial hearing. The Court declared—once again unanimously—that the Obama administration had acted lawlessly. As Justice Alito said, "This kind of thing can't happen in the United States."

It is astonishing that President Obama's conception of federal power is so vast that four liberal justices on the Supreme Court have joined their colleagues in unanimously ruling against the president more than 20 times in five and a half years—double his predecessor's rate of unanimous defeats and 25 percent greater than President Clinton's.

———

But President Obama has not been content merely to stretch the boundaries of executive power according to the Constitution—he has also tried to fundamentally change the document. I refer here to the Democrats' effort to repeal the First Amendment.

Surely, that is hyperbole, you may be thinking. I assure you, it is not.

In the entire time I have been in the Senate, nothing has disturbed me more than the Democrats' efforts to remove the free speech protections found in the Bill of Rights, which provides in the First Amendment that "Congress shall make no law . . . abridging the freedom of speech."

Senate Democrats introduced an amendment to the Constitution that would have amended the Bill of Rights to give Congress broad authority to regulate political speech. Sadly, tragically, astonishingly, every single Democrat in the Senate voted to repeal the free speech provisions of the First Amendment.

Under the terms of the revised First Amendment proposed by the Democrats, Congress would have the authority to "regulate and set reasonable limits on the raising and spending of money by candidates *and others* to influence elections."[10]

Those "others" included every one of us. For example, if a little old lady spent five dollars on cardboard and a stick to make a yard sign for her front yard, Congress would have the power under this amendment to regulate her speech.

But that's not all. A second proposed constitutional amendment would also single out "corporations" and give Congress special power to "prohibit" them "from spending money to influence elections." Now, "corporations" doesn't just include IBM or Exxon Mobil. The NRA is a corporation. So is the Sierra Club, Planned Parenthood, and the Brady Center to Prevent Gun Violence. Just about every citizens group, on the left or right, is organized as a corporation. Under the text of this amendment, Congress could "prohibit" any of them from speaking about politics.

Every presidential election season, *Saturday Night Live* lampoons each party's candidates. Some of that ridicule sticks. There is little doubt that President Ford's image was tarnished by Chevy Chase's wickedly funny impersonation, and that Governor Sarah Palin's was similarly affected by Tina Fey's unforgettable characterization of her. But if Congress can "prohibit" corporations "from spending money to influence elections," then Congress can "prohibit" NBC—a corporation—from airing *Saturday Night Live*.

To be sure, the amendment exempted the news media, specially carving out the "freedom of the press." Thus the *New York Times* would have the ability to speak about politics, but your and my speech would be subject to regulation by Congress. And *SNL* isn't deemed news media, so they fall out of the exemption.

At the end of a hearing of the Judiciary Committee's Subcommittee on the Constitution and Human Rights (on which I then served as the ranking member) on the repeal of the First Amendment, I asked the chairman, Democrat Dick Durbin of Illinois, three simple questions. Do you believe Congress should have the constitutional authority to ban movies? Do you believe Congress should have the constitutional authority to ban books? Do you believe Congress should have the constitutional authority to ban the NAACP (a corporation) from speaking about government?

Senator Durbin hastily gaveled the hearing closed and made his way out of the hearing room without answering those questions. But as we continued the debate weeks later before the full committee, it became obvious that the Democrats' answer to all three questions was yes. President Obama and the Senate Democrats supported amending the Bill of Rights to give Congress the power to ban movies, books, and political speech by political advocacy organizations.

In fact, the Obama administration has already tried to censor a movie, and it has already asserted that it should have the power to ban books. When a conservative group made a movie about Hillary Clinton (*Hillary: The Movie*), the Obama administration tried to fine the moviemaker for daring to make a movie critical of Clinton. The group's name was Citizens United.

Citizens United sued on the grounds that the group was exercising its right to free speech under the First Amendment, and the case went to the Supreme Court. During its defense of censoring Citizens United's movie, the Obama administration was asked at an oral argument if it could also prohibit a company from using its general treasury funds to publish a book that discusses the American political system for five hundred pages and then, at the end, says "Vote for X." President Obama's lawyer said, flat out, "Yes." [11]

These radical claims for the authority to ban books and movies led me to dub the amendment's proponents the "Fahrenheit 451 Democrats," after Ray Bradbury's dystopian classic about book-burning government power run amok. And their positions were so radical that they lost one of their usual allies—the ACLU. To its credit, the ACLU blasted the proposed constitutional amendment: "Even proponents of the amendment have acknowledged that this authority could extend to books, television shows or movies, such as Hillary Clinton's *Hard Choices* or a show like the *West Wing*, which depicted a heroic Democratic presidential administration during the crucial election years of 2000 and 2004." [12]

For the months that Democrats were trying to push their politically driven revision of the Constitution it ought to have been the lead story on the six o'clock news every single night.

Fifty-six Democratic senators proposed repealing the free-speech protections of the First Amendment. The media were silent, and yet the American people deserve to know that a majority of their senators were so afraid of political opposition that they wanted to empower the federal government to police who can speak, and when, and for how long—and on what subjects. That was a radical break from the vision of our Founders.

———

On May 14, 2013, the inspector general of the Treasury Department published a report that described how the Internal Revenue

Service had improperly targeted tea party groups, pro-Israel groups, and pro-life groups based on their conservative political beliefs. In other words, serious allegations were raised that the IRS had been transformed into a political ally of the Obama reelection campaign and the Democratic National Committee.

The day the news became public, the president declared that he was "outraged," he was "angry," and the American people had "a right to be angry" as well.

And yet, in the many months since May 2013, not a single person has been indicted. Many of the victims have not even been interviewed. And Lois Lerner, the head of the rogue office that illegally targeted conservative groups, has gone before the House of Representatives and twice pleaded the Fifth Amendment.

In multiple hearings before the Senate Judiciary Committee, Attorney General Eric Holder obfuscated and stonewalled on the IRS. Under his tenure, the IRS investigation was assigned to a partisan Democrat—a major donor who has given more than six thousand dollars to President Obama and the Democrats. Unsurprisingly, the investigation has gone nowhere.

The fourth estate, despite its overwhelming support for the president and his policies, has also been the target of harassment. Two years ago, without bothering even to reveal its reasons, the administration secretly collected two months of phone records from the staff of the Associated Press, which called the action "a massive and unprecedented intrusion into how news organizations gather news." [13] That same year, they targeted James Rosen, a reporter at Fox News, by labeling him a possible "co-conspirator" in a leak investigation. To quote the *New York Times* editorial board—not something I expect to make a habit of—the Obama administration, with its abuse of Rosen, "moved beyond protecting government secrets to threatening fundamental freedoms of the press to gather news." [14]

In our history presidents of both parties have at times abused their power and exceeded their constitutional authority. But in the past, members of the president's own party—in Congress, in his cabinet, and among independent groups—have shown the courage and principle to stand up to him.

What is unprecedented is the remarkable silence of Democrats in the face of Barack Obama's lawlessness and massive expansion of federal power. The sad fact is that for the Democrats in Washington—and for far too many in my own party as well—politics comes before principle. Electoral considerations come before country. And no offense perpetrated by a party's leader is too outrageous for them to defend.

When President Bush exceeded his constitutional authority and attempted to order the state courts to obey the World Court, I was proud to go before the U.S. Supreme Court on behalf of Texas and defend the Constitution. The Court struck down his unconstitutional order, 6–3. Where are the Democrats willing to do the same to stop their party's abuse of power?

An executive's defiance of the rule of law ought to trouble every American—if only because Barack Obama will not be president forever. Even if you agree with Obama's policies, if this president has the power to ignore the law, then so do his successors—including successors from the opposing party.

This is not what our Founders hoped for. This is not the vision for our country that millions of Americans share. And it does not have to be this way. The genius of the Constitution is that it protects our country from executive branch excesses through the system of checks and balances. The legislative and judicial branches can impose limits on the executive's assertion of power—provided we as public officials have the political will to do it. With the proper leadership, we can restore the purpose and vision behind the American experiment.

CHAPTER 11

★

Reigniting the
Promise of America

A single vote can topple a government, even among the most mighty
on earth. That thought crossed more than one mind on March 28,
1979, as the tallies were being read in the House of Commons. By
a vote of 311 to 310, the Labour government of Prime Minister Jim
Callaghan had fallen. It had lost the confidence of a bare majority,
and would be forced to call for new elections. The outcome was the
first dissolution of a British government on a vote of "no confidence"
in more than five decades.

Among the jubilant members of Parliament that day was the
Leader of the Opposition, a forthright (some might say bracing)
Conservative whose staunch and uncompromising attacks on communism had led her to be dubbed "the Iron Lady" in the Soviet
press. It was a nickname that stuck.

For her part, Margaret Thatcher tried to look inscrutable. But
it was hard to mask her pleasure. She was poised to become the
next prime minister in the upcoming general election, and the first

woman ever to hold the venerable post once occupied by Winston Churchill.

"Hooray!" boomed a voice from the back of the room. The voice belonged to her husband, Denis, whose out-of-order outburst met a quick rebuke from the sergeants at arms (and quite properly so, his wife would later point out).[1]

But Denis shared the excitement of so many conservatives across the nation. A few years earlier, their leader had been on the verge of political extinction. As education minister in a Tory government, she had pushed for cuts to the school lunch program, which included free milk for schoolchildren. For her it was a principled position—we all must tighten our belts—but it led to a furious outcry, even among her fellow Tories. Critics dubbed her "Margaret Thatcher, milk snatcher." From that painful experience, Thatcher learned a valuable lesson: "I had incurred the maximum of political odium for the minimum of political benefit." From then on, she knew when to compromise and when to hold firm. But even as she learned the art of compromise, her conservative principles grew ever stronger. She was studying economics at a think tank founded by a follower of Friedrich Hayek. She had become the public face of opposition to Keynesian economics and the growth of the welfare state.

Thatcher freely shared her disdain for her country's policies, put forward by both major parties, which she believed had led the once-mighty empire down its current road to ruin. The litany was sobering—the nation's industrial supremacy was eroding; union bosses were forming what she called "cartels that restricted competition and reduced efficiency"; coffers and energy were exhausted by involvement in two protracted wars; and the British were suffering from economic and financial anemia. Government was large and unwieldy. In short, Great Britain was a nation "that had had the stuffing knocked out of it."[2]

The Tories—the British equivalent of the Republican Party—
were part of the problem, Thatcher maintained, and had been for
decades. As she later put it, "Almost every post-war Tory victory had
been won on slogans such as 'Britain Strong and Free' or 'Set the
People Free.' But in the fine print of policy, and especially in government, the Tory Party merely pitched camp in the long march to
the left. It never seriously tried to reverse it."[3] Even her own tenure
in government, under Conservative prime minister Ted Heath, had
proven a stunning disappointment—a government, a supposedly
conservative government, which through price controls and efforts
to give coercive power to labor unions had, in her judgment, "proposed and almost implemented the most radical form of socialism
ever contemplated by an elected British government."

To Thatcher, the key to her victory was obvious, fundamental,
and simple. "First you win the argument, then you win the vote."
She knew the socialist foundations of the Labour Party would expose
themselves eventually. But she would not simply wait for the other
party to collapse under its own failings; the damage to Britain in the
meantime might be irreparable. Instead, she resolved to make the case
for the Conservatives under the banner of ideas and principles—less
government, lower taxes, more economic and personal freedom. She
would tell the truth—even if that meant offending members of her
own party who'd grown comfortable during their time in government.

That night, after a small family celebration, Margaret Thatcher
thought of the challenging campaign ahead, the possibility of
failure—not just of the Conservative Party, but of the country she
loved. But she was confident she had made her case to the people;
the die had been cast. And she had a peaceful night's rest.*

---

*The Conservative Party went on that year to win the first of four consecutive elections,
with the country seeing the largest swing in votes from the Labour Party to the Conservatives since 1945.

———

Barack Obama achieved a remarkable thing in 2012: He was re-elected with 51 percent of the popular vote—a higher margin than any Republican presidential candidate has received in twenty-six years. Republicans were for the most part stunned. Our candidate, Mitt Romney, was smart, experienced, a good man—and anointed by the party establishment. But despite the weaknesses in the Obama record, Romney could not make his case. In the words of Lady Thatcher, we didn't "win the argument." Heck, oftentimes we didn't even make the argument.

If nothing else, President Obama's second term has served as a wake-up call that there is a powerful argument to make. Electing, or reelecting, a weak, liberal president does profound damage to a nation. In 1980, we avoided that fate when the Republican standard-bearer declared it was "time to check and reverse the growth of government." He argued without qualification that "lower tax rates mean greater freedom, and whenever we lower the tax rates, our entire nation is better off." He spoke of the pitfalls inherent in "rationing scarcity rather than creating plenty," of "entrepreneurs" as "forgotten heroes," and of the truth that "free enterprise has done more to reduce poverty than all the government programs dreamed up by Democrats."

In 1980, that Republican leader won 91 percent of the electoral college. In 1984, when it was "morning again in America," he won 58 percent of the popular vote and almost all of the electoral college, carrying 49 states. And in 1989, when he retired from public life, he had led not only an administration, but a revolution—the Reagan Revolution.

There will never be another Reagan Revolution, because there will never be another Ronald Reagan. But there can again be Morning in America.

I am convinced that we can reignite the miracle of this promised land of opportunity that God placed between two great oceans—a place where no man is compelled to bow to another, as monarchs and nobles required of generations of peasants and slaves; a place where pilgrims of every era can determine their own destinies, as my immigrant father did; and a place where no agent of any government can take your property, narrow your dreams, or stand between you and your destiny.

To do so, we must together remind ourselves that the promise of America is vital, it is real, and it has the power to transform our lives just as it transformed our parents' and grandparents' lives. Three parts comprise the promise of America.

First, it was born out of the revolutionary principles laid out by our Founders that all men are created equal, and that all are endowed by their Creator with certain unalienable rights to life, liberty, and the pursuit of happiness. These were transformational concepts. For millennia, men and women had been told that our rights come from government, from kings and queens who dole them out like crumbs from their table. America was built on a different proposition: that our rights come from God Almighty. That legitimate government exists only by consent of the governed, and that sovereignty resides not with a monarch, but with We the People. And our Constitution serves, as Thomas Jefferson put it, as "chains to bind the mischief of government."

Second is the incredible opportunity that America has given for each of us to achieve our dreams. Over the last 239 years the audacious premise of this nation—that the People, not the government, know best—has resulted in the greatest engine of opportunity the world has ever known. Generation after generation has flocked to our country in pursuit of the American dream. Of course opportunity is not a guarantee of success, but time and time again, Americans have achieved our aspirations in everything from starting a small business to owning a farm to raising a family to making a difference in thousands of lives. My own family's story is that of risk and reward and loss and renewal—the same journey that continues to beckon as we move further into the twenty-first century fueled by new technologies that hold limitless promise.

And third is American exceptionalism. Our remarkable fusion of political and economic freedom has given America a unique position in the globe. The phrase "American exceptionalism" has been much misunderstood: President Obama famously said in 2009 that he believed in it, "just as the British believe in British exceptionalism or Greeks believe in Greek exceptionalism." But with all due respect to our British and Greek friends, American exceptionalism is different. We are the leader of the free world. The indispensable nation, the country that sets the example for the rest of the world. That doesn't mean that we impose our model on other nations, but that we set the aspirations of what free men and women can achieve. The economic mobility we enjoy has had a global impact, lifting millions out of poverty. Equally remarkably, we have used the greatest military the planet has ever seen to liberate rather than subjugate. America leads. From the Boys of Pont du Hoc battling the evil of the Nazis to the many unsung heroes of the Cold War who fought the Soviets, America has been an extraordinary—and unique—force for good.

The American Constitution, the American dream, American exceptionalism—these are the enduring foundations on which our ongoing national experiment is based. And the good news is that they are permanent things; they are not bound by time or circumstance but are as applicable today as they were in 1776. But just because they are permanent does not mean they are inevitable, and it is our challenge now to defend them so they hold true for our children as well.

Today, a growing number of Americans believe that the promise is receding. In 2014, Pew polling found that 65 percent of Americans believe that our children will be worse off than we are. That is unprecedented. Never before, in more than four centuries of our nation's history, has a majority of Americans not believed that our kids would have a better future. Indeed, from our founding, that's been the American ideal: that our children would have better lives than we did, and their children would have better lives than they did. That ideal is in real jeopardy.

This brings us back around to the challenges that face us in 2015, and the urgent need to reverse the insidious expansion of government power, spending, and debt that has occurred over the last half century, and accelerated dramatically during the Obama administration. This trend threatens to undermine our basic commitment to limited government and liberty and to restrict America's global role, and it makes it imperative that we get back to the commonsense principles upon which this nation was built.

So, how do we reignite the promise of America?

The answer is threefold. First, we must bring back jobs, growth, and opportunity. Economic growth is foundational to every other challenge we face. With growth, we can solve our debt and deficits, lift people from poverty, preserve and reform our entitlements, and rebuild our military strength; without growth, we can do none of those things. Since World War II, our nation has averaged 3.3 percent growth a year. There are only two postwar four-year periods where growth has averaged less than 1 percent: 1978–82 (coming out of the Carter years) and 2008–2012.

History shows a clear cause and effect for jobs and growth. Every time we have pursued out-of-control spending, debt, taxes, and regulation, the result has been economic stagnation and malaise. Conversely, every time we have pursued tax reform and regulatory reform—in the 1920s, in the 1960s, and in the 1980s—the result has been booming economic growth.

What are the policies that will bring back growth? Repealing Obamacare. Reining in abusive regulations. Stopping the EPA from strangling the American energy renaissance that can create millions of high-paying jobs, in energy and in heavy manufacturing. Sound money, auditing the Federal Reserve and stopping its endless quantitative easing that is debasing our currency and making daily life more expensive for hardworking Americans.

And fundamental tax reform. The best tax reform? A simple flat tax, that is fair to everyone. So that everyone can fill out their taxes on a postcard. And, critically, so we can abolish the IRS. To be sure, that's bold, but we are capable of bold accomplishments when the American people get behind them. Two decades ago, Steve Forbes began to build the case for a flat tax, and now—with the political weaponization of the IRS under the Obama administration—conditions are stronger than ever for fundamental reform.

If, after the last recession, the United States had enjoyed the rate of economic growth it has enjoyed in an average recovery, American families would have about $10,000 more income per family than they now do. But because President Obama imposed unprecedented taxes and regulations, the United States is right now missing around a *trillion* dollars in expected productivity.[4]

We should set an audacious goal of enacting policies to encourage the private sector to create 10 million new jobs. Enough for full recovery. Good, blue-collar jobs with strong wages and work with dignity. High-paying white-collar jobs in expanding technologies. Full-time jobs, not people trapped in endless part-time positions. Multiple, exciting job opportunities for young people coming out of school. Get government out of the way and unleash the creativity of millions of small businesses.

Abolishing the IRS will also weaken the power of career politicians, and the ability of Washington to frustrate the will of the American people. So too will passing structural reforms, like a constitutional Balanced Budget Amendment, term limits for Congress, and a lifetime ban on former members of Congress ever lobbying.

Second, we need to protect our constitutional rights. Our founding charter has served us well for more than two centuries. It protects liberty by separating powers, limiting the authority of the federal government, and guaranteeing every American the freedom to speak your mind, pray to God, and protect yourself and your family by bearing arms in their defense. Every single one of those constitutional protections has come under assault from the Obama administration, which has usurped the power of Congress through executive amnesty, redefined the relationship between the federal government and the governed through Obamacare, and attempted to repeal and undermine the First and Second Amendments through abusive campaign finance regulations, coercions of religious consciences, and repeated attacks on the right to bear arms.

We need to vigorously protect free speech and religious liberty, and make clear that the federal government has no authority to undermine the Second Amendment. Likewise, we need to protect the Fourth and Fifth Amendment privacy rights of every American. We need to defend life, from conception to natural death. And we need to respect the Tenth Amendment, the fundamental limitation on the authority of the federal government.

The Tenth Amendment makes clear that there are a host of issues that should be decided by the states. Issues like marriage. Rather than the federal government or federal courts trying to impose new definitions of marriage, it should be left where the matter has been decided for two centuries: in the hands of democratically elected state legislatures. Personally, I strongly support traditional marriage between one man and one woman. A covenant ordained by God. But if people want to try to change the legal standards of civil marriage, the proper way to do so is to convince their fellow citizens. It is not for unelected judges to tear down the traditional marriage laws adopted by the people.

Likewise, education is primarily a matter for the states. The federal government has no business trying to determine curricula in our schools. Education is far too important for it to be dictated from Washington; ideally, it should be at the local level, where parents have direct control over the education of their kids. For that reason, we should repeal Common Core, and make clear that matters of curricula are outside the authority of unelected federal bureaucrats.

Third, we must restore American leadership in the world. Reigniting the promise of America means more than defending freedom at home; it requires the defense of American interests abroad. Here again, President Reagan is a strong example. Viewed from the perspective of the Obama era—in which Russia annexed Crimea and ISIS seized much of Iraq—it is remarkable that in Reagan's nearly three thousand days in office, not a single inch of ground fell to the communists. How did he accomplish that? Well, he did *not* begin his presidency with an apology tour. He did *not* draw red lines that he later ignored. And he did *not* betray our allies, reward our enemies, or "lead from behind." As Reagan said after leaving the White House, his foreign policy showed "the sky would not fall if America restored her strength and resolve. The sky would not fall if an American president spoke the truth. The only thing that would fall is the Berlin Wall."

In 2013, I was proud to attend the funeral of Nelson Mandela—the only senator to attend, and one of only two Republicans. I admired Mandela because he was a freedom fighter, he stood up to racial injustice and transformed his nation and the world. But, when his odious supporter Raul Castro spoke at the funeral, I walked out of the stadium.

"Tear down this wall" changed history, and we need to return to being a clarion voice for freedom. We should be calling evil by its name and speaking out for the unjustly oppressed, whether it is Pastor Saeed Abedini in Iran or Meriam Ibrahim (now freed from) Sudan or Leopoldo López in Venezuela.

At the same time, America has always been reluctant to engage in military conflict, and we should show humility in our foreign policy. It is not the job of our military to try to democratize every country on earth, or to turn Iraq into Switzerland.

It is, however, the job of the military to protect our nation, and we need to rebuild and modernize our armed forces to do so. And we cannot hide from the hard work of defending our national security. That means standing by our friends and allies and standing up to our enemies when needed. If and when military force is required, it should begin with a clearly defined objective, directly tied to our national security interests. We should use overwhelming force, and then we should get the heck out.

Peace through strength means that when our enemies believe America will act, it often is unnecessary to do so. Weakness, however, of the Obama-Clinton-Kerry variety, only invites more aggression and escalates the chances of military conflict.

So, if we understand what to do substantively to turn our country around, how do we get it accomplished? How do we actually win at the ballot box?

The secret to GOP victory in 2016 really isn't much of one. It is, in fact, obvious to those who are willing to learn from the past. The most consistent pattern of the last forty years is that Republicans win the White House whenever we nominate a candidate who runs as a strong, principled conservative with a positive, optimistic, hopeful message. We lose whenever we nominate the "more electable" candidate who runs as a mushy establishment moderate.

History is on the side of conservatives on this point. In 1968 and again in 1972, Richard Nixon ran for the presidency as a strong "law and order" conservative. He didn't always govern as one, but that was how he ran for office—highlighting clear contrasts between his positions and those of his liberal opponents, Hubert Humphrey and George McGovern. The contrast between the conservative Nixon and the liberal McGovern was so great in 1972 that Nixon won forty-nine states.

In 1976, Ford ran as an establishment moderate, and lost.

In 1980 and 1984, Ronald Reagan, perhaps the most unapologetically conservative candidate since Calvin Coolidge, won two straight elections, defeating Carter's reelection bid and then winning a forty-nine-state landslide in 1984.

The following election, in the year 1988, offers the most compelling example in the entire litany. Reagan's vice president, George H. W. Bush, ran as a strong conservative. As, in effect, the third term of Reagan. He won in a landslide. Then, in 1992, Bush had moved to the middle. He had violated his pledge not to raise taxes. He put a liberal justice, David Souter, on the Supreme Court. In style and substance, he publicly distanced himself from the Reagan years. Bush 41 is a good and decent man, but, running for reelection as the "electable" establishment moderate, he lost, in the process giving the nation two terms of Bill Clinton.

In 1996, Bob Dole—another courageous war hero—likewise ran as a moderate, and lost.

Then, in 2000 and 2004, George W. Bush ran for office as a principled Reaganite and won both times.

After that, we were back to the candidacies of the "electable"

John McCain in 2008 and Mitt Romney in 2012, both honorable men running hard to the middle.

The Romney loss, the most recent defeat for the GOP on the presidential level, really stung because it should not have happened. By the end of President Obama's second term, our economy was on the ropes. The rich were richer and the poor a whole lot poorer, notwithstanding Obama's rhetoric about fixing the income inequality that reached new heights under his policies. Labor force participation had plummeted, and nearly 90 million Americans were not working. Our international standing was an embarrassment. Our commander in chief was a textbook study in indecision and vacillation.

Governor Romney is a good man who ran hard, but he failed to "win the argument," as Thatcher would put it. Indeed, the entire 2012 race can be summed up in two words: 47 percent. Romney's infamous gaffe arose after a liberal surreptitiously gained entry into a private fund-raising event and taped Romney's off-the-cuff remarks to donors. The particular comment that sparked outrage was the following:

> There are forty-seven percent of the people who will vote for the president no matter what. All right, there are forty-seven percent who are with him, who are dependent upon government, who believe that they are victims, who believe the government has a responsibility to care for them, who believe that they are entitled to health care, to food, to housing, to you-name-it. . . . *And so my job is not to worry about those people*—I'll never convince them that they should take personal responsibility and care for their lives.

Romney's initial observation, that the Democrats are using entitlements to try to make voters dependent on the big government, has some substantive force. But his conclusion—*"my job is not to worry about those people"*—is precisely backward. Rather than writing off the 47 percent, we need to reach out to them with the economic opportunity that has been our greatest advantage as a country.

I believe—no, I know—that the vast majority of our citizens desperately want to stand on their own two feet. To know the dignity of work. No grandmother, no *abuela* in Texas, wants her children or grandchildren to subsist on welfare. Americans want to provide for their families—to give them a shot at a better life than we ourselves have had.

For too long, the left has gotten away with the lie that Republicans are the party of the rich—even as the rich get richer under big-government policies and a large percentage of the richest Americans give generously to the Democrats in every election cycle. But Republicans play into that theme when they seem to write off a huge swath of the country—and refuse to make the case for our beliefs and ideals. Nobody is going to vote for you if they believe you don't like them. Nobody is going to vote for you if you don't even talk to them.

I recognize that Romney's comment was likely a verbal slip, but it became the overarching narrative of the last election. Designating any group of Americans—let alone almost half our population—as helpless dependents is directly contrary to our values as a nation.

That is not how conservatives think. We understand that you cannot build a lasting governing coalition by telling nearly half the country that you don't care about them.

Conservatives should—and must—champion the Americans who've been forgotten and left behind. After nearly seven years of President Obama, there are more of them than ever before. They see a skyrocketing stock market on the one hand and stagnant real wage growth on the other. They see the Washington insiders getting fatter while real families struggle to get by. And they are, in some ways, the nation's foremost experts on the limits of liberalism—because they are paying the price for liberalism's empty promises.

When Reagan ran for president, he campaigned in cities destroyed by liberal policies and he spoke to Americans who'd likely never before looked a Republican candidate for any office in the eye. In Detroit, he said, "More than anything else, I want my candidacy to unify our country, to renew the American spirit and sense of purpose. I want to carry our message to every American, regardless of party affiliation, who is a member of the community of shared values." Four months later, he carried Michigan by six and a half percentage points—a quarter of a million votes.

Reagan won by offering a hopeful message, an aspirational message. He told us to "always remember that you are Americans, and it is your birthright to dream great dreams in this sweet and blessed land, truly the greatest, freest, strongest nation on earth." He explained what it meant to be a conservative and how those policies could help more people become part of the American dream—regardless of their race or religion or political affiliation. As a result, he built a coalition of voters, many of them blue-collar workers who usually voted for the other party, and in the process coined a new term: Reagan Democrats. The conservative Reagan is the only president in modern history who has a group from the other party named after him.

If the consultants were right, that running to the middle earns you crossovers, then there would be Ford Democrats, Dole Democrats, McCain Democrats, and Romney Democrats. There are not. And it's easy to see why: If you're a Democrat, and the two candidates are close ideologically, what are you going to do? You'll stick to your team and vote for the Democrat. But, in 1980, the choice was stark: Reagan drew a line in the sand, and millions of voters who had been FDR Democrats said, "Those are my values." That's how you win crossovers.

Why do I look to the example of Reagan so often? Many reasons. One is simply admiration and respect; his leadership transformed the world. But two, the times are very similar. The situation today is very much like the late 1970s. Indeed, the parallels between Jimmy Carter and Barack Obama are uncanny. Same failed economic policies, same stagnation and malaise. Same feckless and naïve foreign policy; indeed, the very same countries—Russia and Iran—openly laughing at and mocking the president of the United States.

I believe 2016 will be an election very much like 1980. To win, we have to paint in bold colors, not pale pastels.

In 1980, Ronald Reagan was seen as a remarkably divisive figure—within the Republican Party; indeed, he had just challenged, and nearly beaten, Gerald Ford, the sitting Republican president in the Republican primary. Within Washington, Reagan was despised. That's because Reagan fundamentally rejected the accepted establishment wisdom of how to win an election. He didn't abandon his beliefs and run to the middle. He explained his beliefs and brought the middle to him.

When you paint in bold colors, two big things happen: You turn out the base, by the millions; and you earn more crossover votes. The latter consequence seems counterintuitive, and it's directly contrary to the narrative of the Washington consultants.

In 2012, the Obama campaign very shrewdly understood their circumstances. They knew that the Reagan Democrats—the Ohio steelworkers—would not be voting for the president's reelection. An Ohio steelworker's life has been made incredibly difficult under the Obama economy. So their goal was to keep as many of them home as possible. The Obama team launched saturation attack ads to paint Mitt Romney as a rich, out-of-touch elitist—and the 47 percent line fit their strategy perfectly. The Ohio steelworkers (and Reagan Democrats nationally) stayed home. Evangelical Christians, whom the Romney campaign apparently just assumed would vote for him, stayed home, too.

This point bears underscoring. If we want to win, we need to be clear-eyed and data driven. And if you look to the data, if you compare 2004 (the last race that Republicans won nationwide) to 2008 and 2012, the biggest difference is the millions of conservatives who stayed home. They fall primarily into two categories—evangelical Christians and Reagan Democrats. Millions of them stayed home in 2008, and even more did so in 2012.

The only way to win in 2016 is to bring back the conservatives who are staying home. And if we nominate another candidate like Bob Dole or John McCain or Mitt Romney—all good, honorable men, but all lost—then the same voters who stayed home in 2008 and 2012 . . . will stay home again in 2016. And Hillary Clinton will be the next president.

That cannot happen.

Instead, we must reassemble the Reagan coalition. Conservatives, libertarians, evangelicals, young people, Hispanics, African-Americans, women, Jewish voters, Reagan Democrats—we need to unite behind our shared values.

Part of uniting that coalition is just telling the truth, not just telling one group what they want to hear, and another group what they want to hear. There are values and principles that stitch us together. Every one of us believes in the American dream, in growth and jobs and opportunity and in individual liberty and constitutional freedoms. Those are themes and principles that cut across race and class and gender. Those were the themes that are at the heart of any true grassroots campaign.

Let's dispense with divisive, interest group politics, and ignore the empty happy talk that consultants tell us we need to reach the middle. There is an alternate course, what I've labeled "opportunity conservatism." The fact is that conservative policies and beliefs are aspirational. We should look with the single-minded focus of easing people's ascent up the economic ladder. And that applies to everyone. In other words, Republicans shouldn't disparage the 47 percent. We should embrace them. It is the Republican Party that is and should be the party of the 47 percent, of the poor, the up-and-coming, the struggling middle class, and those who want a better circumstance for themselves and a better one than that for their children.

To the hardworking men and women who want to believe again in the promise of America, our movement must give bold voice and action to reigniting the unlimited potential of every American.

The simplest principle behind opportunity conservatism is the aphorism we all know: "Give a man a fish, it feeds him for a day; teach a man to fish, it will feed him for a lifetime." We need to be defending the opportunity to take responsibility for our own lives because that is what has consistently led to extraordinary prosperity and achievement.

Opportunity conservatism rejects the wealth-redistribution policies of the left as a means for upward mobility. Among other problems, collectivist approaches—punitive tax policies on corporations or upper-income families—simply do not work. They fail to produce economic prosperity or to improve the material conditions of the populace. And they lead to bankruptcy and economic collapse, as Europe demonstrates daily. Widespread economic redistribution places enormous burdens on small businesses, kills jobs, and rarely helps the recipients of government largesse. Even now in the United States, such high-tax, collectivist approaches are failing to produce results for the people these policies are ostensibly trying to help. Under the Obama administration, for example, the unemployment rate climbed above 10 percent among Hispanics in 2012 and to 14 percent among African-Americans.

Whenever entrepreneurs and small businesses suffer, those struggling to improve their economic conditions are hurt the worst. Yet Republicans rarely talk about this. It is as if we've read too much of the opposition's talking points, as if we are too afraid to make the simple, unarguable point that free-market policies expand opportunity, produce prosperity, and improve lives—especially for those working to climb the economic ladder.

For centuries, our free enterprise system has been the path to the American dream. I know this is not an abstract, academic theory. Like all of us, I've seen it in my own family. When my dad earned fifty cents an hour, he was filled with hope for a better life. Better than imprisonment and torture in Cuba, better than struggling for food each day. He believed, as Ronald Reagan once said, that "in America, our origins matter less than our destinations."

I am a conservative today because I firmly believe that free-market policies—which produce a robust and growing small business environment—are the best policies to lift every American, including teenage immigrants like my dad was, up to prosperity.

My family's story is like the stories of countless other families who have the same hopes and dreams for their children that my mom and dad had for me. This is in a sense every family's love story with America. Today, for example, 2.3 million Hispanics—roughly one in every eight Hispanic households—own small businesses, trying to get their piece of the American dream. The reason that so many millions have come from all over the world to America is that they recognize that no other nation on earth offers such opportunity. No other nation allows a citizen to come here with nothing and achieve anything.

This is the advantage that the Republican Party has over its opposition. The Democrats are the party of government; of dependency. They have no choice in the matter. Their constituencies are big labor union bosses that depend on government largesse and special interest organizations that survive by encouraging a victim mentality among their supporters.

Republicans, by contrast, can champion policies of self-sufficiency, responsibility, and economic mobility. We believe our potential should never be limited by our government, but only by our talent and imagination. As Reagan said at his first inauguration, our mission is to make government "work with us, not over us; to stand by our side, not ride on our back . . . provide opportunity, not smother it; foster productivity, not stifle it." Every issue we debate and discuss on the nation level should be framed this way—as issues of whether government is facilitating dependence or self-reliance; enslavement to the state or individual freedom.

Conservatives are against excessive governmental regulations, but rarely explain why. The reason is that these regulations kill jobs and restrict Americans' ability to achieve, earn a living, build a business, and buy a first home.

Conservatives support lower taxes and a fundamental reform of the tax code so that it is fairer and simpler. But we need to explain why: because when we lower taxes, we give more money to entrepreneurs and business owners to build their businesses and create jobs. We give more money to heads of households so that they can make their own financial decisions for their children and grandchildren. When we reform the tax code, we give more power to families and less power to accountants and lawyers who are enriched by a tax code that only they, and their buddies in Congress, can understand.

Conservatives are wary of big union bosses. We need to explain why: because unions confiscate wages to fill their own coffers and pursue their own agenda. By demanding costly regulations on growing businesses, the union bosses make it harder for low-skilled workers to get jobs.

Conservatives favor educational reform, such as vouchers and scholarships and charter schools. Again we need to explain why: because education reform empowers parents and expands opportunities for kids struggling to get ahead in schools that have failed them. It is at its core a civil rights issue, and it is fundamentally unfair to trap kids in bad schools because of their race, ethnicity, income level, or simply because they live in the wrong zip codes.

Conservatives favor Social Security reform and personal retirement accounts. This allows low-income Americans to accumulate wealth on their own and pass it on to their children and grandchildren. Giving more power, more money, and more control to Americans allows them to make their own decisions and realize the American dream. As Reagan said, "individual freedom and the profit motive were the engines of progress which transformed an American wilderness into an economic dynamo that provided the American people with a standard of living that is still the envy of the world."

When we make such commonsense arguments, critics in the media and on the left—as if those were two separate entities—invariably attack us for advocating "selfishness." Self-responsibility is somehow morphed into self-interest. That's how distorted the thinking in Washington is—giving people more of their own money is considered selfish. Allowing people to have more responsibility and more control over their own lives is considered reckless.

There is perhaps no better response to these arguments than those once famously uttered by one of my heroes, Milton Friedman. Confronted in the 1970s by a liberal talk-show host about the "greed" behind capitalism, he responded with a smile, saying:

> "Well first of all, tell me: Is there some society you know that doesn't run on greed? You think Russia doesn't run on greed? You think China doesn't run on greed? The world runs on individuals pursuing their separate interests. The

great achievements of civilization have not come from government bureaus. Einstein didn't construct his theory under order from a bureaucrat. Henry Ford didn't revolutionize the automobile industry that way. In the only cases in which the masses have escaped from the kind of grinding poverty you're talking about, the only cases in recorded history, are where they have had capitalism and largely free trade. If you want to know where the masses are worse off, worst off, it's exactly in the kinds of societies that depart from that. So that the record of history is absolutely crystal clear, that there is no alternative way so far discovered of improving the lot of the ordinary people that can hold a candle to the productive activities that are unleashed by the free-enterprise system."

One of the things that encourages me is the new generation of leaders in the Republican Party who are stepping forward. Virtually all of them are clustered within just a few years of each other: Mike Lee (43), Marco Rubio (43), Paul Ryan (45), Nikki Haley (43), Tim Scott (49), Rand Paul (52), Bobby Jindal (45), Joni Ernst (44), Tom Cotton (37), and Mia Love (39).

All of us were kids when Ronald Reagan was president. I'll go to my grave with Ronald Wilson Reagan defining what it means to be president. The World War II generation referred to FDR as "our" president. Reagan, I believe, made an indelible mark on this new generation of leaders. In fact, I've called this generation the "children of Reagan." If you listen to Marco or Tim or Nikki, the language they use is positive, optimistic, unifying, appealing to our better angels. It's not the mean, nasty, divisive, wedge issues that have characterized so many Republicans. It echoes Morning in America. The way we win is to stand for principle, but to do so in a way that paints a brighter future for our nation.

Finally, reigniting the promise of America requires courage. Just as it took courage for Margaret Thatcher to challenge the "good old boys" of her party's establishment, and just as it took courage for Reagan to run against an incumbent president of his own party in 1976, conservatives must stand up to the establishment impulses of our own party, and it will take courage for all of us to do so. Is there a political risk to doing everything possible to repeal Obamacare? Yes. Are there political dangers involved in fighting to stop the president's unconstitutional, lawless amnesty? You bet. But if standing up for conservative values were easy, we wouldn't have needed a tea party movement of millions of men and women standing up to Washington politicians in both parties.

I believe in the promise of our nation, and if you're reading this, I suspect you do, too. It is our turn to step up and defend it, because as my father always used to say to me, "If we lose our freedom here, where will we go?"

We have great friends and allies around the world, but if the last seven years have taught us anything it is that an absent America, an America that thinks global leadership means voting "present" at the United Nations, leads to chaos. Our promise is powerful, but it requires vigilance to defend it.

This is our fight. And it's not enough for conservative leaders to just say they're leading. They need to show it with action, not with words. They need to stand with you, rather than only asking you to stand with them. And they need to listen to you, rather than lecturing you—because only when elected representatives in Washington start listening to the American people will we reignite the promise of America.

ACKNOWLEDGMENTS

Many hands were involved in the crafting of this book, and I am grateful to have this opportunity to acknowledge them. That list starts, of course, with my wife, Heidi, who went on much of this adventure with me, as well as our daughters, Caroline and Catherine, who made that journey all the more meaningful and joyous.

In many ways this book is a celebration of two Americans of grit, passion, and fortitude. I am as ever very proud and thankful to be the son of Rafael Cruz and Eleanor Darragh.

I'm grateful for the editorial advice and reminiscences of my dear friends David Panton and Chad Sweet. I also appreciate the guidance of those on my staff who assisted with this effort. Victoria Coates took time from an already busy schedule to help oversee this project from start to finish. I thank as well my friends and advisors John Drogin, Jason Johnson, Jason Miller, Jeff Roe, Chip Roy, Alec Aramanda, and Scott Keller. In helping to cull together a diffuse collection of photographs, my thanks to Samantha Leahy, Josh Perry, Bruce Redden, Bobby Rodriguez, and David Sawyer.

As a new author, I have benefited greatly from the skill and persistence of my agents, Keith Urbahn and Matt Latimer of Javelin. I also owe a debt of thanks to an accomplished lawyer and fierce advocate for her clients, Cleta Mitchell.

Adam Bellow and Eric Meyers led the charge for this book at HarperCollins on a very quick turnaround schedule. Their advice and guidance is greatly appreciated.

I owe a profound and eternal debt to the people of the great state of Texas for having faith in me, for giving me the honor of representing them in our nation's capital, and for offering a steady and reliable supply of support and prayers. Those prayers have strengthened me enormously during our many ongoing battles in Washington, D.C.

Lastly, and most importantly, I offer my gratitude to the American people and to our blessed and wonderful country. Every day of my life I remember the lessons of my father, a Cuban immigrant who came here with nothing. Only in a land like America is his story—is our story—even possible.

NOTES

------------- ★ -------------

Introduction

1. Matthew Yglesias, "It's Worth Actually Reading Obama's 2006 Debt Ceiling Speech," Slate.com, October 9, 2013, http://www.slate.com/blogs /moneybox/2013/10/09/obama_s_2006_debt_ceiling_speech.html.
2. Ibid.
3. Stephen Dinan, "U.S. Borrows 46 Cents of Every Dollar It Spends," *Washington Times*, December 12, 2012, http://www.washingtontimes.com/news /2012/dec/7/government-borrows-46-cents-every-dollar-it-spends/.
4. Alexandra Jaffe, "McConnell Sets Up Fight Over Debt Limit," *The Hill*, January 26, 2014, http://thehill.com/blogs/congress-blog/economy-budget /196434-mcconnell-sets-up-fight-over-debt-limit.
5. Betsy Woodruff, "Behind Closed Doors, a Messy Fight over the Debt-Ceiling Hike," *National Review Online*, February 12, 2014, http://www .nationalreview.com/corner/370987/behind-closed-doors-messy-fight-over -debt-ceiling-hike-betsy-woodruff.
6. Allahpundit, "Revealed: Senate clerks didn't announce names during debt-ceiling vote so that Republicans could secretly switch," HotAir.com, February 13, 2013, http://hotair.com/archives/2014/02/13/revealed-senate -clerks-didnt-announce-names-during-debt-ceiling-vote-so-that-republi cans-could-secretly-switch/.
7. Manu Raju, "Some GOP Colleagues Angry with Ted Cruz," Politico, October 2, 2013, http://www.politico.com/story/2013/10/ted-cruz-blasted -by-angry-gop-colleagues-government-shutdown-97753.html.
8. "The Minority Maker," *Wall Street Journal*, February 12, 2014, http:// www.wsj.com/articles/SB1000142405270230443410457937937428735 7650.

Chapter 1

1. "Cuba dissident Farinas awarded Sakharov Prize by EU," BBC News, October 21, 2010, http://www.bbc.com/news/world-europe-11594804.

Chapter 2

1. James L. Haley, *Sam Houston* (Tulsa: University of Oklahoma Press, 2004), 390.

Chapter 3

1. http://news.yahoo.com/u-generals-sex-crimes-trial-delayed-indefinitely-151003096.html.

Chapter 4

1. http://www.nytimes.com/2001/08/10/us/execution-approaches-in-a-most-rare-murder-case.html.
2. Court documents can be found here: http://www.clarkprosecutor.org/html/death/US/beazley779.htm.
3. CNN Transcript, "Scheduled to Die," May 25, 2002, http://edition.cnn.com/TRANSCRIPTS/0205/25/cp.00.html.

Chapter 6

1. Barry Goldwater, *With No Apologies: The Personal and Political Memoirs of United States Senator Barry M. Goldwater* (New York: William Morrow, 1979), 262.
2. Ibid.
3. Stanley Kutler, *The Wars of Watergate* (New York: Knopf, 1990), 532.
4. Goldwater, *With No Apologies*, 262.
5. Kutler, *The Wars of Watergate*, 535.
6. Goldwater, *With No Apologies*, 261.
7. Ibid., 262.
8. Bob Woodward and Carl Bernstein, *The Final Days* (New York: Simon & Schuster, 1976), 398.
9. Goldwater, *With No Apologies*, 263.
10. Ibid.
11. Kutler, *The Wars of Watergate*, 542.
12. Ibid.

13. Ibid.
14. Goldwater, *With No Apologies*, 263.
15. Kutler, *The Wars of Watergate*, 539.
16. Goldwater, *With No Apologies*, 261.
17. Woodward and Bernstein, *The Final Days*, 414.
18. Goldwater, *With No Apologies*, 267.
19. Woodward and Bernstein, *The Final Days*, 414.
20. Ibid.
21. Ibid., 415.
22. Ibid.
23. Ibid.
24. Ibid.
25. Goldwater, *With No Apologies*, 268.
26. http://www.texnews.com/1998/2003/texas/texas_Maverick_929.html.
27. Michael King, "Maps, Balls and Testifying," *Austin Chronicle*, December 26, 2003, http://www.austinchronicle.com/news/2003-12-26/191173/.
28. Compare http://clerk.house.gov/member_info/electionInfo/2002/2002 Stat.htm#43 with http://clerk.house.gov/member_info/electionInfo/2006 /2006Stat.htm#43.
29. Medellín's confession is available here: http://www.americanbar.org/con tent/dam/aba/publishing/preview/publiced_preview_briefs_pdfs_07_08 _06_984_RespondentAppendix.authcheckdam.pdf.
30. Allan Turner, "Medellín executed for rape, murder of Houston teens," *Houston Chronicle*, August 5, 2008, http://www.chron.com/news/houston -texas/article/Medellín-executed-for-rape-murder-of-Houston-1770696 .php.

Chapter 7

1. "Cover Story: Ronald for Real," *Time*, October 7, 1966, http://content.time .com/time/magazine/article/0,9171,199339,00.html.

Chapter 8

1. PolitiFact, "Ted Cruz says Americans invented 'Pong,' 'Space Invaders' and the iPhone," June 6, 2009. http://www.politifact.com/texas/statements /2013/jun/06/ted-cruz/ted-cruz-says-americans-invented-pong-space-in vade/.

Chapter 9

1. Quotes from Reagan's 1961 record on socialized medicine come from Eric Zorn, "Ronald Reagan on Medicare, circa 1961. Prescient rhetoric or familiar alarmist claptrap?" *Chicago Tribune*, September 2, 2009, http://blogs .chicagotribune.com/news_columnists_ezorn/2009/09/ronald-reagan -on-medicare-circa-1961-prescient-rhetoric-or-familiar-alarmist-claptrap -.html.

2. Lenin's quote was likely not as pithy. Various accounts dispute the authenticity and translation of the quote, but it appears that Lenin did express the idea in his private notes.

Chapter 10

1. http://www.theatlantic.com/politics/archive/2009/06/-brotherhood-in vited-to-obama-speech-by-us/18693/.

2. Jonathan Turley, "Turley: Obama The President That Richard Nixon Always Wanted To Be," Real Clear Politics, June 3, 2014, http://www.real clearpolitics.com/video/2014/06/03/turley_obama_the_president_that _richard_nixon_always_wanted_to_be.html.

3. For a complete list, see Ted Cruz, "The Legal Limit: The Obama Administration's Attempts to Expand Federal Power," http://www.cruz.senate.gov /files/documents/The%20Legal%20Limit/The%20Legal%20Limit%20 Report%204.pdf.

4. Barack Obama, "Remarks by the President on the Government Shutdown," October 3, 2013, http://www.whitehouse.gov/the-press-office/2013/10/03 /remarks-president-government-shutdown.

5. Barack Obama, "Remarks by the President to the National Council of La Raza," http://www.whitehouse.gov/the-press-office/2011/07/25/remarks -president-national-council-la-raza.

6. President Barack Obama, "Remarks on Immigration Reform," Google Hangout, February 21, 2013. https://www.youtube.com/watch?v=e9lmy _8FZM.

7. Barack Obama, "Obama: 'If Congress Won't Act, I Will,'" ABC News.com, October 29, 2011, http://abcnews.go.com/Politics/video/obama-congress -wont-act-14841368.

8. *United States v. Jones*; *Sackett v. EPA*; *Hosanna-Tabor Evangelical Lutheran Church & School v. EEOC*; *Arizona v. United States*; *Gabelli v. SEC*; *Arkansas Fish & Game Commission v. United States*; *PPL Corp. v. Commissioner of Internal Revenue*; *Horne v. USDA*; *Sekhar v. United States*; *NLRB v. Noel Canning*; *Riley v. California*; *Bond v. United States*; *Burrage v. United States*;

Judulang v. Holder; *United States v. Tinklenberg*; *Henderson ex rel. Henderson v. Shinseki*; *Carachuri-Rosendo v. Holder*; *United States v. O'Brien*; *Abuelhawa v. United States*; *Flores-Figueroa v. United States*. All twenty decisions are listed and explained in Senator Ted Cruz, "The Supreme Court Has Unanimously Rejected the Obama Administration's Arguments 20 Times," http://www.cruz.senate.gov/files/documents/The%20Legal%20 Limit/Report_5.pdf.

9. *Riley* concerned the actions of state police rather than federal agents, but the Obama administration's theory extended to federal power. *Riley v. California*, Justia U.S. Supreme Court, June 25, 2014, https://supreme.justia.com /cases/federal/us/573/13-132/.

10. U.S. Senate, 113th Congress, *S.J. Res. 19, A Joint Resolution Proposing an Amendment to the Constitution of the United States Relating to Contributions and Expenditures Intended to Affect Elections*, January 11, 2014, https:// www.congress.gov/bill/113th-congress/senate-joint-resolution/19/all-info.

11. *Citizens United v. Federal Election Commission*, No. 08-205, 2009, http:// www.supremecourt.gov/oral_arguments/argument_transcripts/08-205 .pdf.

12. Laura W. Murphy, "The ACLU Supports Campaign Finance Reform and Free Speech," ACLU.org, August 12, 2014, https://www.aclu.org/blog/ free-speech/aclu-supports-campaign-finance-reform-and-free-speech.

13. Laura Sellers-Earl, "APME Condemns Justice Department Actions," Associated Press Media Editors, May 14, 2013, http://www.apme.com/news /125392/APME-condemns-Justice-Department-actions.htm.

14. Editorial, "Another Chilling Leak Investigation," *New York Times*, May 21, 2013, http://www.nytimes.com/2013/05/22/opinion/another -chilling-leak-investigation.html?_r=1.

Chapter 11

1. Margaret Thatcher, *The Downing Street Years* (New York: HarperCollins, 1993), 3.

2. Ibid., 5.

3. http://www.independent.co.uk/voices/comment/our-hate-figures-and-he roes-are-mere-surfers-on-the-tide-of-history-8572433.html.

4. Stephen Moore, "Obama's Illusory Economic Recovery," *Washington Times*, January 25, 2015, http://www.washingtontimes.com/news/2015 /jan/25/stephen-moore-obamas-illusory-economic-recovery/print/.

INDEX

★

ABOUT THE AUTHOR

—————————— ★ ——————————

In 2012, Ted Cruz was elected the thirty-fourth U.S. Senator from Texas. A passionate fighter for limited government, economic growth, and the Constitution, Ted won a decisive victory in both the Republican primary and the general election, despite having never before been elected to office. Before joining the Senate, he was the solicitor general of Texas. Ted and his wife, Heidi, live in his hometown of Houston, Texas, with their two young daughters, Caroline and Catherine.